ACS SYMPOSIUM SERIES **498**

Biocatalysis at Extreme Temperatures
Enzyme Systems Near and Above 100 °C

Michael W. W. Adams, EDITOR
University of Georgia

Robert M. Kelly, EDITOR
North Carolina State University

Developed from a symposium sponsored
by the Division of Biochemical Technology
at the 201st National Meeting
of the American Chemical Society,
Atlanta, Georgia,
April 14–19, 1991

American Chemical Society, Washington, DC 1992

Library of Congress Cataloging-in-Publication Data

Biocatalysis at extreme temperatures: enzyme systems near and above
  100 °C / Michael W. W. Adams, editor; Robert M. Kelly, editor.

  p.   cm.—(ACS symposium series, ISSN 0097–6156; 498)

"Developed from a symposium sponsored by the Division of
Biochemical Technology at the 201st National Meeting of the American
Chemical Society, Atlanta, Georgia, April 14–19, 1991."

Includes bibliographical references and index.

ISBN 0–8412–2458–7

1. Microbial enzymes—Thermal properties—Congresses. 2. Thermophilic bacteria—Congresses.

I. Adams, Michael W. W., 1954– . II. Kelly, Robert M., 1953– .
III. American Chemical Society. Division of Biochemical Technology.
IV. American Chemical Society. Meeting (201st: 1991: Atlanta, Ga.)
V. Series.

QR90.B55 1992
576′.11925—dc20   92–16399
                                                          CIP

The paper used in this publication meets the minimum requirements of American National
Standard for Information Sciences—Permanence of Paper for Printed Library Materials, ANSI
Z39.48–1984.   ∞

ACS Symposium Series

QR 90
B 55
1992
CHEM

M. Joan Comstock, *Series Editor*

1992 ACS Books Advisory Board

Foreword

THE ACS SYMPOSIUM SERIES was founded in 1974 to provide a medium for publishing symposia quickly in book form. The format of the Series parallels that of the continuing ADVANCES IN CHEMISTRY SERIES except that, in order to save time, the papers are not typeset, but are reproduced as they are submitted by the authors in camera-ready form. Papers are reviewed under the supervision of the editors with the assistance of the Advisory Board and are selected to maintain the integrity of the symposia. Both reviews and reports of research are acceptable, because symposia may embrace both types of presentation. However, verbatim reproductions of previously published papers are not accepted.

Contents

INDEXES

Preface

WHEN APPROACHED BY PROFESSOR PRASAD DHURJHATI of the University of Delaware to organize a symposium on Biocatalysis Near and Above 100 °C, we were faced with a dilemma. While this area had aroused its share of scientific and technological curiosity, it was not clear that enough had been accomplished to serve as a basis for such a symposium. Most efforts had focused on the ecology of geothermal environments and the growth physiology of hyperthermophilic microorganisms. The pioneering work of individuals such as Holger Jannasch of Woods Hole Oceanographic Institute, Karl Stetter of the University of Regensberg, and John Baross of the University of Washington had led to the discovery of a growing list of novel microorganisms, many of which have optimal growth temperatures at or above 100 °C. However, the purification and characterization of the enzymes from these organisms had only begun, largely due to the difficulties encountered in working with biological systems at such high temperatures. As such, biocatalysis at high temperatures remains on the fringes of biology and biotechnology.

Nonetheless, a number of interesting studies concerned enzymes derived from hyperthermophiles, many of which are still in progress. The preliminary results from many of these studies are intriguing and potentially have great impact beyond the ecology and microbial physiology of geothermal environments. Already molecular biology has been expanded through the use of thermostable DNA polymerases in the polymerase chain reaction (PCR), and thermostable enzymes have been employed in the bioprocessing of starch. However, the enzymes employed in these applications, while stable relative to most enzymes, have limited durability at temperatures approaching and exceeding 100 °C. Some enzymes from hyperthermophiles (bacteria with optimal growth temperatures near or above 100 °C) are known to have half-lives that are orders of magnitude longer than those of the enzymes currently used. However, the extent to which these hyperthermophilic enzymes can be utilized remains to be seen.

This book contains contributions from a diverse set of research efforts that are focused in some way on biocatalysis at elevated temperatures. In organizing this volume, we attempted to provide coverage ranging from microbiology to the molecular biology and biotechnology of high-temperature enzymes. Much to our delight, contributors were willing to present and discuss recent results, helping to make the perspective presented here forward-looking and optimistic. Whether these enzymes are used directly in biotechnology or information gained from their study can be extended to improve our understanding of biocatalysis, this research area will undoubtedly make its mark in many scientific areas. We hope that this first symposium that seeks to address these issues will lead to others with similar intentions.

We thank the contributors to this volume for their willingness and cooperation. Their frank discussion of ongoing research projects is appreciated. We also are grateful to Barbara Tansill of ACS for her encouragement and patience. Finally, our thanks to the U.S. Department of Energy, the National Science Foundation, the National Institutes of Health, and the Office of Naval Research, whose support has fostered this emerging research area.

MICHAEL W. W. ADAMS
University of Georgia
Athens, GA 30602–7229

ROBERT M. KELLY
North Carolina State University
Raleigh, NC 27695–7905

January 1992

Chapter 1

Biocatalysis Near and Above 100 °C

An Overview

Michael W. W. Adams[1] and Robert M. Kelly[2]

[1]Department of Biochemistry and Center for Metalloenzyme Studies,
University of Georgia, Athens, GA 30602
[2]Department of Chemical Engineering, North Carolina State University,
Raleigh, NC 27695–7905

Enzymes are typically labile molecules and thus are adversely affected when exposed to any type of extreme conditions. As such, biocatalysis, in either a physiological or biotechnological sense, has usually constrained to a rather narrow range of temperature, pH, pressure, ionic strength and to an aqueous environment. In fact, given its physiological role, and the need at times to regulate enzymatic activity, this is appropriate. Unfortunately, the use of biological catalysts for technological purpose necessitates that enzymes be stable and functional in non-physiological environments. The challenge then is to either isolate enzymes more suitable for a particular application or be able to modify existing enzymes systematically to improve their stability and/or function. While a number of thermostable enzymes have been studied previously, the focus here is on thermostable enzymes produced by high temperature microorganisms.

It is now a fact that life can be found at temperatures at or above the normal boiling point of water. Bacteria have been isolated from sites as diverse as the ocean floor and terrestrial hot springs proliferating at these temperatures. As the number of microorganisms isolated from geothermal environments grows, there is increasing interest in the intrinsic characteristics of their constituent biomolecules. The enzymes from these bacteria are not only significant for their potential as biocatalysts but as model systems to which nature has endowed incredible levels of thermostability. It will be up to scientists and technologists interested in the properties of these proteins to devise ways to probe the most fundamental aspects as well as envision ways in which biocatalysis at elevated temperatures can be put to good use.

The prospect of biocatalysis at high temperatures and in normally denaturing environments is not a new one. Studies focusing

0097–6156/92/0498–0001$06.00/0

on ribonuclease A and on several α-amylases that are stable at temperatures in the vicinity of 100°C have shown that enzymes from mesophilic or moderately thermophilic sources may possess high levels of thermostability. In fact, there have been numerous attempts to systematically improve the thermostability of a given protein through site-directed mutagenesis or to infer the bases for thermostability by comparing amino acid sequences of homologous systems. However, while the idea of biocatalysis near and above 100°C may not be new, the levels of thermostability at these temperatures found in enzymes from high temperature bacteria exceed any that have previously been examined. Previous work on enzymology at high temperatures focused on enzymes that, from a thermostability standpoint, were uncommon. Now, the expanding sources of high temperature bacteria provide a diverse source of extremely thermostable enzymes for both basic scientific studies as well as for biotechnological opportunities.

There is no doubt that the study and use of enzymes from extremophiles is in its earliest stages. Although prospects from both basic and applied perspectives are appealing, only a relatively small number of researchers have significant efforts in this emerging field. This is not surprising since one must develop a range of capabilities including biomass generation in order to make significant progress in reasonable periods of time. Fortunately, there are an increasing number of efforts along these lines which makes progress on many fronts increasingly likely.

Despite the fact that the study of enzymes from high temperature bacteria is not well-developed, there has, nonetheless, been considerable progress on many associated issues. This volume contains a diverse set of contributions that, taken together, provide some perspective of where this field is now and where it is heading. As is clear from examining the diverse expertise collected here, many different approaches and motivations are apparent. These range from investigations into the fundamental and molecular characteristics of high temperature enzymes to their physiological significance to ways in which they can be utilized for technological purpose. A brief discussion of these contributions is in order.

Chapter 2 provides a perspective on the genesis of efforts on biocatalysis at elevated temperatures. There are still only a relatively small number of microorganisms with extremely high optimal growth temperatures populating only a handful of genera. The initial efforts focusing on the biochemistry of the enzymes from these bacteria were motivated by an interest in comparative physiology and biochemistry relative to mesophilic counterparts. Chapter 3 provides some perspective on the challenges facing one interested in this emerging area of science and technology. Two case studies featuring enzymes from *Pyrococcus furiosus* illustrate the many differences one encounters in studying biocatalysis at elevated temperatures. Chapter 4 further demonstrates the process through which high temperature enzymes are identified and isolated with a discussion of thermostable ureases from organisms in hot springs at Yellowstone National Park.

Much of the work done thus far in this area has emerged from a interest in comparative physiology between mesophiles and thermophiles. The enzymes discussed in Chapters 5 and 6 illustrate this point. Chapter 5 focuses on the respiratory enzymes implicated in the physiological function of *Pyrodictium brockii*. Not only does this chapter provide an interesting perspective on sulfur respiration but demonstrates an efficient use of very small amounts of biological material - a critical requirement in many cases in this field. Chapter 6 focuses on glutamate dehydrogenase in *P. furiosus* which apparently is under regulatory control as part of the bacterium's nitrogen metabolism.

Chapters 7 and 8 discuss the effect of additional stresses on enzymes from high temperature bacteria. Chapter 7 reviews the status of enzymology in organic solvents and includes coverage of prospects for catalyzing reactions in non-aqueous media and at elevated temperatures. The prospect of organic synthesis in this environment is presented. Many high temperature bacteria come from locations of extremely high pressure; their enzymes must therefore deal with this additional stress. Chapter 8 provides some background on pressure effects on biocatalysis, including the additional affect of temperature.

It is difficult with such limited information to begin to generalize as to the mechanisms of extreme protein thermostability. However, it is possible to extend present knowledge of protein stability to attempt to provide some basis for further study. Chapters 9 through 11 come at this in different ways. Chapter 9 uses insights developed from theoretical studies of protein thermostability to provide a thermodynamic basis for higher levels of thermostability. While meaningful experimental data are not available to test the current hypothesis, the arguments presented are persuasive and merit additional attention. Chapter 10 presents a correlative approach to predict enzyme properties from very limited information. This framework could be very useful is assessing the relative merits of particular enzymes in a given situation. Chapter 11 then describes the ambitious task of deriving insights into the intrinsic features of thermostability using molecular dynamics.

Possibly the most exciting development in biotechnology in recent years has been the use of the polymerase chain reaction to amplify quantities of DNA to useful levels. A thermostable DNA polymerase is central to this technology and several have now been isolated from high temperature bacteria. A more fundamental question arises as to how these organisms stabilize/destabilize DNA at such high temperatures. Chapter 12 provides some intriguing insights into how the replication process proceeds at elevated temperatures and the role of DNA-binding proteins in this regard. From a technological perspective, Chapter 13 discusses a new DNA polymerase isolated from *P. furiosus* and its properties relative to other enzymes now in use for the same purpose.

In one sense this volume provides no definitive answers to the question of what underlies extreme thermostability or how biocatalysis proceeds at high temperatures. However, the contributions here do provide a perspective on the field and suggest the directions in which it is headed. It is clear that progress has been made and, because of the potential benefits to be derived from studies of enzymes from high temperatures bacteria, more attention to this field is merited.

RECEIVED March 3, 1992

Chapter 2

Metabolic Enzymes from Sulfur-Dependent, Extremely Thermophilic Organisms

Michael W. W. Adams[1], Jae-Bum Park[1], S. Mukund[1], J. Blamey[1], and Robert M. Kelly[2]

[1]Department of Biochemistry and Center for Metalloenzyme Studies, University of Georgia, Athens, GA 30602
[2]Department of Chemical Engineering, North Carolina State University, Raleigh, NC 27695–7905

Microorganisms growing near and above 100°C were discovered only in the last decade. Most of them depend upon elemental sulfur for growth. Both the organisms and their enzymes have enormous potential in both basic and applied research. To date only a few metabolic enzymes have been characterized. The majority are from two sulfur-dependent organisms, from the archaeon, *Pyrococcus furiosus,* and from *Thermotoga maritima,* the most thermophilic bacterium currently known. In this chapter we review the nature of the sulfur-dependent organisms, their evolutionary significance, and the properties of the enzymes that have been purified so far.

The existence of life forms that not only survive but actually thrive at temperatures near and even above the normal boiling point of water is a very recent discovery in biology. The field began in 1982 when Karl Stetter of the University of Regensburg isolated microorganisms from shallow marine volcanic vents off the coast of Italy that grew reproducibly above 100°C (*1*). This immediately led, with considerable success, to the search for organisms with similar properties. However, an understanding of how biochemical processes are able to take place at such temperatures has been less forthcoming. Indeed, it was only in 1989 that the first metabolic enzymes were purified and characterized from an organism capable of growing at 100°C (*2,3*). Similarly, the first insights into the factors that lead to protein "hyperthermostability" are only just emerging (Chapter 11), as are potential mechanisms for stabilizing DNA (Chapter 12). The main objectives of this chapter are to describe the "extremely thermophilic" organisms known at present, and to summarize the properties of the enzymes and proteins that have been purified to date from the sulfur-dependent species.

0097–6156/92/0498–0004$06.00/0

The Classification of Extreme Thermophilic Organisms

As shown in Figure 1, well over a dozen different genera are now known that are able to grow optimally at temperatures of 80°C or above (*4-7*). These are referred to as extremely thermophilic or hyperthermophilic organisms, although the latter term is frequently restricted to those organisms that grow optimally at or above 100°C. As indicated, all but one of the extreme thermophiles have been classified within what is now termed the *Archaea* domain of life. These contrast with the numerous thermophilic organisms belonging to the *Bacteria* domain that have been isolated over the years, most of which have temperature optima for growth (T_{opt}) below 70°C (*8*).

The relationship between the *Archaea* and the *Bacteria* is shown in Figure 2. During the 1960's, two forms of life were recognized based on features at the cellular level. One was the *Eucaryotae,* which represented all "higher" life forms, and these were distinguished by the presence of a membrane-defined nucleus. The other was the *Procaryotae,* which included all bacteria, and these lacked a nucleus. During the 1970's techniques were developed to classify organisms at the molecular level using sequences of proteins and nucleic acids. A fundamental dichotomy in the procaryotes was then shown to exist by Woese and coworkers on the basis of 16S rRNA sequences (*9, 10*). This confirmed in quantitative terms the close relatedness of the the vast majority of bacteria (procaryotes) and the fundamental difference between them and the eucaryotes. However, two groups of organisms, methanogenic bacteria and extremely halophilic bacteria, were shown to be specifically related to each other, but widely separated from all other bacteria and also from eucaryotes. The organisms in the methanogen/halophile branch were termed archaebacteria to distinguish them from all other bacteria, or eubacteria. This separation was also reflected in the biochemistry of the two groups: the archaebacteria have different cell walls and cell membranes, and contain a range of enzymes and cofactors not present in other bacteria. Indeed, the genetic machinery of the archaebacteria in many ways more resembles that of eucaryotes rather than eubacteria.

As more molecular sequences became available, it was possible to discern the relationship between these three domains of life, and a universal phylogenetic tree was proposed by Woese and colleagues in 1990 (*11*). In fact, they recommended that the three domains of life be referred to as *Bacteria, Archaea,* and *Eucarya,* since, at the molecular level, the archaebacteria resembled other (eu)bacteria no more than they did eucarya(otes), and so they should no longer be termed bacteria. These definitions will be used herein.

A rather surprizing conclusion from the universal tree shown in Figure 2 is that the *Eucarya* and *Archaea* have a common ancestor that is not shared by the *Bacteria.* Moreover, the first organisms to have diverged from the *Eucarya/Archaea* lineage were the extremely thermophilic archaea. Thus, investigations into the molecular properties of the extreme thermophiles may give some insights into the development of the eucaryotic cell. Furthermore, as shown in Figure 1, there is one extremely thermophilic genus that is not classified within

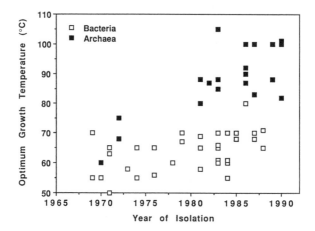

Figure 1. Thermophilic organisms isolated over the last twenty years. Data taken from Table I and Refs. 7, 8.

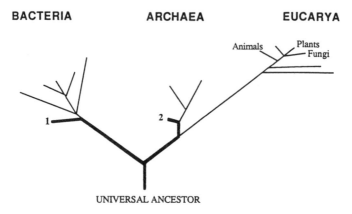

Figure 2. The universal phylogenetic tree showing the three domains of life. Known extremely thermophilic organisms include the novel bacterium, *Thermotoga maritima* (designated 1) and the S°-dependent archaea (designated 2). It is proposed that the extent of extreme thermophily, shown by the heavy line, includes the Universal Ancestor. Evolutionary time is proportional to the distance along any one line. Present time is represented by the ends of the lines. (Adapted from reference 11.)

the *Archaea* domain, and this is *Thermotoga* (*13*). As shown in Figure 2, in addition to being the most thermophilic, *Thermotoga* is also the most ancient bacterial genus currently known, the first to have diverged from the "Universal Ancestor". Extremely thermophilic organisms therefore appear to be the remnant of some ancestral form of all extant life, having evolved when the earth was much hotter than it is at present (*12*). Of course, this is contrary to traditional wisdom, which has regarded thermophily as an adaptation. That is, thermophiles were thought to represent a few "conventional" organisms that originally grew at "normal" temperatures but had somehow adapted so that they could flourish at higher temperatures. However, the remarkable conclusion from the evolutionary analysis is that extreme thermophiles are the most ancient organisms currently known. The rest of the life forms on this planet may therefore themselves be the result of temperature adaptations - adaptations to lower (less than 100°C) temperatures.

Physiology of Extreme Thermophiles

The genera of extremely thermophilic organisms currently known are depicted in Table I. Species of the genera *Pyrococcus, Pyrobaculum, Pyrodictium, Hyperthermus* and *Methanopyrus,* together with ES-4, grow optimally at or above 100°C, and these will be termed "hyperthermophiles". Most of the genera shown in Table I are represented by only one or two species. The majority have been found in geothermally-heated marine environments, in both shallow (several meters below sea level) and deep water (several kilometers below sea level). Two of the organisms shown in Table I, ES-1 and ES-4, have yet to be formally classified. Both were recently isolated by Baross and colleagues near deep sea hydrothermal vents. Stetter and coworkers have also recently extended the known habitats of extremely thermophilic bacteria. They isolated several different bacteria from within the crater and from the open sea plume of an erupting submarine volcano, located 40 m below sea level (*44*). These organisms included relatives (by DNA hybridization) of species of *Pyrodictium, Pyrococcus, Archaeglobus* and *Thermococcus,* bacteria that had been previously found only in volcanic vents off the coast of Italy. They also isolated a novel bacterium that showed no DNA homology with any of the extreme thermophiles tested. Some of the extreme thermophiles therefore appear to be spread in the open oceans, and are able to remain viable even under such cold and aerobic conditions.

Except for three methanogenic genera and one novel sulfate-reducing genus (Table I), all of the extremely thermophilic archaea are termed sulfur-dependent organisms, since, to a greater or lesser extent, they obtain energy for growth by the reduction or oxidation of elemental sulfur (S°). Only species of *Acidianus* and *Desulfurolobus* are able to grow aerobically, and they do so by oxidizing S° with O_2 to produce sulfuric acid. Accordingly, they grow optimally at low pH (near 2). They are also autotrophs and use CO_2 as a carbon source. Remarkably, these organisms are facultative aerobes, and are able to switch to an anaerobic growth mode using H_2 as an electron donor and

Table I. Extremely Thermophilic Genera

Genus	T_{max} (T_{opt})	%GC	Habitat[a]	First Isolated	Ref.
BACTERIA					
Thermotoga	90° (80°)	46	m	1986	13-15
ARCHAEA					
S°-DEPENDENT					
Thermoproteus	92° (88°)	56	c	1981	16,17
Staphylothermus	98° (92°)	35	d/m	1986	4,18
Desulfurococcus	90° (87°)	51	d/m/c	1982	19-21
Thermofilum	100° (88°)	57	c	1983	22
Pyrobaculum	102° (100°)	46	c	1987	23
Acidianus	96° (90°)	31	m/c	1986	24
Desulfurolobus	87° (81°)	33	c	1986	25
Pyrodictium	110° (105°)	62	d/m	1983	7,26,27
Thermodiscus	98° (90°)	49	m	1986	4
Pyrococcus	105° (100°)	38	d/m	1986	28,29
Thermococcus	97° (88°)	57	d/m	1983	30-32
Hyperthermus	110° (100°)	56	m	1990	33
"ES-1"	91° (82°)	59	d/m	1990	34
"ES-4"	108° (100°)	55	d/m	1990	35
SULFATE-REDUCING					
Archaeoglobus	95° (83°)	46	m	1987	36-38
METHANOGENIC					
Methanococcus	86° (85°)	31	d/m	1983	39
Methanothermus	97° (80°)	33	c	1981	40,41
Methanopyrus	110°(100°)	60	d/m	1990	42,43

[a] Species have been isolated from shallow marine vents (m), deep sea hydrothermal vents (d), and/or from continental hot springs (c). The sulfur-dependent genera are grouped in separate orders, except for Hyperthermus, ES-1 and ES-4. Modified from refs. 2 and 4.

S° as the electron acceptor. Phylogenetically, both genera belong to the archaeal order *Sulfolobales,* which includes the less thermophilic (T_{opt} < 80°C) genera, *Sulfolobus* and *Metallosphaera* (*7*).

All of the other S°-dependent archaea are strict anaerobes. Most obtain energy for growth by S° respiration - the reduction of S° to H_2S. Some species of *Thermoproteus* and *Pyrobaculum* grow autotrophically and use H_2 as the electron donor for S° reduction. All the others are strict heterotrophs and use organic substrates to reduce S°. The growth of some of these organisms, including *Hyperthermus butylicus* (*33*) and *Pyrodictium abyssum* (*27*), is also stimulated by the addition of H_2, although why this should be so is completely unknown. In any event, with the three exceptions noted below, optimal growth of all of these organisms is dependent upon S°, and thus the production of H_2S. Some species, such as *Thermococcus celer, Thermodiscus maritimus, Pm. abyssum* and *H. butylicus,* have been reported to grow well in the absence of S° by a fermentative type of metabolism. However, we have found that the growth of all of these organisms is almost insignificant unless S° is added to the media.

The obligate requirement of S° has severely hindered the growth of extremely thermophilic organisms, and the production of biomass in sufficient quantities for the the purification of their enzymes. The production of large amounts of H_2S prohibits the use of conventional stainless steel fermentors because of the corrosion problem, a situation made worse by the relatively high salt concentrations that are required by many of the marine organisms. In addition, accurate temperature control of most conventional fermentors is not possible much above 70°C. For small scale studies on the order of a few liters, glass vessels can be used, and such techniques have been used with considerable success by Maier and colleagues to study the H_2 metabolism of *Pyrodictium brockii* (see Chapter 4). Similarly, Kelly and coworkers recently described a continuous culture system that can operate near 100°C using S°-containing media (*45*) However, for large scale cultures of H_2S-producing extreme thermophiles, corrosion-resistant, enamel-lined fermentation systems are the only alternative. Such large scale systems (300 liters and above) are in operation in Karl Stetter's laboratory at the University of Regensburg and just recently at the author's facility at the University of Georgia.

Only two "S°-dependent" archaea are known at present that appear to grow equally well with and without S°. They are *Pyrococcus furiosus* and *Thermococcus litoralis.* This property is also shared by the one bacterium that can be considered extremely thermophilic, *Thermotoga maritima* (*13*). These organisms are all obligate heterotrophs, and appear to obtain energy for growth by fermentation rather than by S° respiration. Before describing the properties of some of the enzymes that have been purified from them, it is relevant to put the study of extreme thermophiles in perspective.

The Unique Potential of Extreme Thermophiles

Although our knowledge of extremely thermophilic organisms is mainly limited to the papers reporting their isolation, the exploitation of their extraordinary properties for biotechnological purposes can be anticipated in the very near future. Extremely thermophilic bacteria and their thermostable enzymes have numerous biotechnological advantages over their mesophilic counterparts (5,8,46,47), and the notion of hyperthermophilic enzymes presents some intriguing commercial possibilities. They may provide an opportunity to bridge the gap between biochemistry and a great deal of industrial chemistry. However, as shown in Table I, all of the extremely thermophilic and hyperthermophilic $S°$-dependent bacteria were isolated in the 80's, and most in just the last few years. Hence the few biochemical characterizations reported with these organisms have focused mainly on their genetic machinery, e.g. RNA's and polymerases, to try and gain insight into their relationship with the rest of the microbial world (4,7,12,48,49). Consequently, little is known about the metabolisms of these organisms, and even less about the enzymes involved. Three unique research opportunities are therefore presented by these bacteria, and particularly with the hyperthermophilic species: a) elucidating the novel metabolic pathways that are likely to be present in these remarkable organisms, b) determining the mechanisms by which these organisms stabilize biomolecules, especially proteins, and c) understanding how their enzymes are specifically adapted to catalyze reactions at extreme temperatures. Let us briefly consider all three aspects.

The $S°$-dependent extreme and hyperthermophiles mainly represent the third major branch of the *Archaea* domain, and independent of temperature, might be expected to be a source of novel and unexpected biochemistry. The unusual natural environments of these organisms may also be a factor. For example, hydrothermal vent fluids contain minerals such as iron and manganese, and gases such as methane, CO_2 and H_2, at concentrations many orders of magnitude above those in normal sea water. Thus, the availability of both nutrients and minerals may have severely limited the metabolic options of extreme thermophiles, or may have enabled them to develop novel catalytic systems *because* of the availability of a particular nutrient, not normally found in more conventional ecosystems. Are the metabolic pathways of the hyperthermophiles minor variants of those of mesophilic organisms, or do completely different pathways exist ? Coupled to this, of course, are the unique biochemical aspects of life at extreme temperatures. Chemical reactions obviously take place at 100°C that do not readily occur at ambient temperatures, gases are much less soluble, and many "universal" biomolecules, such as NAD, CoA, and ATP, are extremely unstable above 90°C (50). How have the hyperthermophiles circumvented, or even taken advantage of, such basic problems ? As described below, at least one hyperthermophile, *P. furiosus,* may metabolize glucose to acetate without the participation of the thermolabile cofactor, NAD(P). Moreover, this pathway is dependent upon tungsten, a rarely used element in biological systems, but one that could be prevalent in deep sea hydrothermal vent fluids.

Elucidation of all of the factors which contribute to the enhanced thermostability of some proteins has, and continues to be, one of the most challenging problems in both biotechnology and biochemistry, e.g. Refs. 51-53. The availability of hyperthermophilic proteins opens up a completely new aspect of protein stability, especially since it has been shown that proteins *in vitro* may be destroyed at 80°C - 100°C by the hydrolysis of certain peptide bonds, deamination of asparaginyl residues, and cleavage of disulfide bonds (*54*). Significant insight into stabilizing mechanisms should be obtained by comparisons of homologous hyperthermophilic and mesophilic proteins at the molecular level. However, as will be shown below, amino acid sequence information is not sufficient and three-dimensional structural information is required. In addition, although most of the proteins that have been purified from extreme thermophiles are remarkably thermostable, some of them, such as the hydrogenase of *T. maritima* and AOR from *P. furiosus* (see below), are unstable at ambient temperatures. This "cold-lability" may suggest that factors that confer resistance to high temperatures can also lead to instability at low (20°C) temperatures.

The other intriguing and potentially useful property of the extremely thermophilic enzymes is that they have no or very low activity at ambient temperatures. Similarly, some of their metal-containing redox centers cannot be or are only slowly reduced at 20°C, or are only reduced or oxidized by substrates above 80°C. This is in agreement with the idea that thermostable proteins are much more "rigid" at ambient temperatures than mesophilic proteins (*55*), and presumably there is insufficient "flexibility" to allow interactions between adjacent redox centers or between substrates and redox centers. This has enormous utility in that one can, in essence, carry out "cryoenzymology" at and above ambient temperatures, without the complications of cryosolvents. Some of these factors will be considered in the following description of the enzymes and proteins that have been purified so far from two extreme thermophilic organisms. These are the hyperthermophilic archaeon, *Pyrococcus furiosus*, and the novel extremely thermophilic bacterium, *Thermotoga maritima*.

Pyrococcus furiosus

P. furious was isolated from a shallow marine vent by Stetter and coworkers in 1986 (*28*). It is a strictly anaerobic heterotroph, and utilizes both simple (maltose) and complex (starch) carbohydrates for growth with the production of organic acids, CO_2 and H_2 as products. It will not use ammonia or free amino acids as an N-source, instead, proteins or peptides (tryptone or yeast extract) are required. The organism will also grow, albeit poorly, with these complex substrates as the sole N and C source (*28*). The H_2 produced during fermentation inhibits growth, but this can be relieved either by sparging with an inert gas, or by the addition of S°, which is reduced to H_2S. Growth rates and cells yields appear to be similar when either method is used, indicating that S° reduction is not coupled to energy conservation (28). *P. furiosus* grows between approximately 80 and 105°C. At the optimum temperature, 100°C, the doubling time is about 40 min. We routinely obtain cell yields of approximately 800 g (wet weight) from a 400 liter culture (2).

P. furiosus produces an extracellular amylase when starch is used as a growth substrate. This enzyme was recently purified (56), as was the amylase from the related organism, *P. woesii* (57). Both enzymes convert starch to oligosaccharides containing mainly 2 - 5 sugar units. They have M_r values of 136,000 (*P. furiosus*) and 70,000 (*P. woesii*), and both are extremely thermostable: maximal catalytic activity was observed at 100°C and the half-life of the enzymes at 120°C was about 2 hr. Like many other starch-utilizing organisms, *P. furiosus* also produces an intracellular α-glucosidase to further hydrolyze the products of the amylase to glucose. This enzyme was recently purified by Kelly and coworkers (58). It is a monomeric protein of M_r 125,000 with a temperature optimum for catalysis near 110°C. It's half-life at 98°C was about 48 hr. These carbohydrate-metabolizing enzymes from *P. furiosus* are among the most thermostable proteins currently known.

Insight into the pathway for glucose metabolism in *P. furiosus* developed from our finding that tungsten greatly stimulates the growth of this organism (2). A tungsten-containing enzyme was subsequently purified, and this was shown, using spectroscopic and potentiometric techniques, to be a novel oxidoreductase that catalyzed a reaction of extremely low potential (59). Tungsten (W) is an element seldom used in biological systems, indeed, at the time this work was carried out, only one other W-containing enzyme was known. This was a formate dehydrogenase isolated by Ljungdahl and coworkers from the acetogen, *Clostridium thermoaceticum* (60). However, the W-containing protein from *P. furiosus* did not catalyze formate oxidation, and analyses showed that the enzyme as first purified was in fact an inactive form of an aldehyde ferredoxin oxidoreductase, which we now refer to as AOR (61). AOR is very O_2-sensitive and intrinsically unstable at ambient temperatures. It has to be purified rapidly in buffer containing glycerol, dithiothreitol and sodium dithionite under strictly anaerobic conditions to prevent substantial losses of activity. AOR catalyzes the oxidation of a variety of aldehydes to the corresponding acid, a reaction of very low potential ($E_m \sim -520$ mV), but it did not a) oxidize glucose or aldehyde phosphates, b) utilize CoA, and c) reduce NAP or NADP. Instead, it used a redox protein known as ferredoxin as an electron carrier. We had earlier purified this ferredoxin from *P. furiosus* (3) and showed that it was the physiological electron donor for the hydrogenase of this organism, the enzyme responsible for catalyzing H_2 production. This enzyme has also been purified and characterized (2).

Ferredoxin from *P. furiosus* is also a remarkably stable protein, being unaffected by 12 hr at 95°C (3). It has a M_r value of 7,500 and its amino acid sequence is known (Howard, J. B., Park, J.-B. and Adams, M. W. W., unpublished data). Ferredoxins have been purified from a wide range of mesophilic organisms, but the most thermostable of them has a half-life at 80°C of only about 20 min. Over thirty amino acid sequences of mesophilic ferredoxins are known (*62*), but direct comparisons of these with the sequence of the *P. furiosus* protein offer no clue as to why the latter is so much more stable at high temperatures. The electron carrier abilities of these ferredoxins reside with either one or, in most cases, two [4Fe-4S] clusters. The cluster is bound to the protein via its Fe atoms which are coordinated to the sulfur atoms of four cysteinyl residues   The *P. furiosus* protein is unique in this regard in that it is the only example of a 4Fe-ferredoxin in which one of the four cysteinyl residues has been replaced, in this case by an aspartyl residue.   A recent electron nuclear double resonance study indicated that an OH⁻ molecule rather than the aspartate group binds to the unique Fe site (*63*). Because of the lack of complete cysteinyl coordination, the [4Fe-4S] cluster in *P. furiosus* ferredoxin has unusual electron paramagnetic resonance and resonance Raman properties (*3,64*) and it will also bind exogenous ligands such as cyanide (*65*). Similarly, the unique Fe atom can be easily removed to give a [3Fe-4S] cluster, and can be reconstituted with other metal ions (M) to yield mixed metal [MFe_3S_4] clusters, such as the first example of a [$NiFe_3S_4$] cluster in a biological environment (*66*).

P. furiosus* hydrogenase also has some unique properties (*2,6*). Like the majority of hydrogenases that have been purified from mesophilic organisms, e.g. Ref. 67, the enzyme contains a Ni site and Fe/S clusters. However, it differs in that it has high H_2 evolution activity in in vitro assays, it uses ferredoxin as its physiological electron carrier, and it is insensitive to CO, acetylene and nitrite, which are the classical inhibitors of the mesophilic enzymes.   The hydrogenase is also extremely thermostable.   The pure enzyme has a half life at 100°C of about 2 hrs, and the optimum temperature for catalyzing both H_2 evolution and H_2 oxidation is above 95°C.   Interestingly, the hydrogenase evolves H_2 using reduced *P. furiosus* ferredoxin as the electron donor at significant rates only above 80°C, at the growth temperature of the organism.

Since *P. furiosus* produces H_2 during growth, a combination of W-containing AOR, ferredoxin and hydrogenase, all of which are present at quite high concentrations in this organism, could oxidize aldehydes to H_2 and the corresponding acid. However, the conventional pathways for glucose oxidation found in mesophilic organisms do not require an aldehyde oxidation step, so we proposed (*61*) that *P. furiosus* contains a partially non-phosphorylated glycolytic pathway similar to the NAD(P)-dependent pathway thought to be present in some aerobic archaea (*68*).   The pathway in *P. furiosus* differs in that all of the oxidation steps are linked to the reduction of ferredoxin, and thus to H_2 production via the hydrogenase.   As shown in Figure 3, AOR is proposed to catalyze a key step, the oxidation of glyceraldehyde to glycerate.   A unique consequence of the proposed "pyroglycolytic" pathway is that *P. furiosus* could convert glucose to acetate without the participation of the thermolabile cofactors, NAD(H) or NADP(H).   In other words, these cofactors appear to have been replaced by an

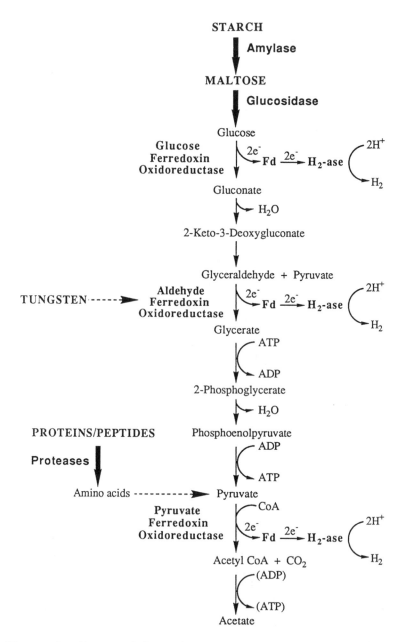

Figure 3. Proposed "pyroglycolytic" pathway in *Pyrococcus furiosus*. Starch, maltose, tungsten and proteins/peptides are growth substrates. Enzymes that have been characterized are shown in bold type. Fd and H_2-ase represent ferredoxin and hydrogenase, respectively. (Adapted from reference 61.)

extremely thermostable protein, ferredoxin. The first oxidation step in the proposed pathway is catalyzed by another new enzyme, glucose ferredoxin oxidoreductase (Figure 3). Like AOR, this enzyme is extremely O_2 sensitive and it rapidly loses activity during multi-step column chromatography of cell-free extracts. It has yet to be obtained in a pure state.

The final oxidation step in the pathway shown in Figure 3, the production of acetyl CoA and CO_2 from pyruvate, is catalyzed by pyruvate ferredoxin oxidoreductase, termed POR. This extremely O_2-sensitive enzyme was recently purified from *P. furiosus* (Blamey, J. and Adams, M. W. W., unpublished data). In contrast to the PORs from mesophilic bacteria, which are dimers of $M_r \sim 250,000$ and require added thiamine pyrophosphate (TPP) for activity (*69*), the *P. furiosus* enzyme is a trimer of M_r 93,000 and its activity is independent of TPP. The optimum temperature for the catalytic activity of *P. furiosus* POR is above 95°C, but unlike the other enzymes from *P. furiosus* described above, it is not very thermostable: it has a half-life at 90°C of about 20 min. How *P. furiosus* generates ATP from pyruvate was recently investigated by Schäfer and Schönheit (*70*). They identified an acetyl CoA synthetase (ADP forming) in cell-free extracts, which produced ATP and acetate from acetyl CoA, ADP and phosphate.

The pathway in Figure 3 therefore accounts for the utilization of starch, maltose, and tungsten by *P. furiosus,* and for the production of H_2, CO_2 and acetate. We now turn to the N metabolism of this organism, about which very little is known. As already mentioned, proteins or peptides serve as N sources, and extremely active and extremely thermostable proteases have been identified in *P. furiosus.* Kelly and coworkers (*71*) made the remarkable discovery that upon boiling cell-free extracts with sodium dodecyl sulfate (SDS, 1%) for 24 hr, only four proteins could be detected and two of these had proteolytic activity. These were termed the S66 and S102 proteases, which reflected their M_r values of 66,000 and 102,000, respectively. S66 was shown to be a serine-type protease that had a half-life at 98°C of over 33 hr. Three other proteases that were not resistant to SDS treatment were also identified in cell-free extracts. Similar results were recently obtained by de Vos and colleagues, who termed the serine protease-like activity "pyrolysin" (*72*). Presumably, the amino acids generated by these proteolytic activities can be used directly for biosynthetic purposes, and as shown in Fig. 2, can provide some energy via transamination reactions by entering the pathway at the level of pyruvate. However, as yet, none of the enzymes involved in amino acid metabolism in *P. furiosus* have been characterized, with the exception of glutamate dehydrogenase, which uses both NAD and NADP as electron carriers (*73*: see Chapter 6).

P. furiosus exhibits significant growth when tungsten is not added to the growth medium (*2*). Thus, either there is sufficient contaminating tungsten in the medium to support this growth via the proposed pathway (Figure 3), or the organism may contain a more conventional though less efficient means of oxidizing carbohydrates. The latter may well be the case as low amounts of glyceraldehyde-3-

phosphate dehydrogenase (GAPDH), triose phosphate isomerase, and phosphoglycerate kinase have been found in cell-free extracts (61). Moreover, GAPDH has been purified from the related organism, *Pyrococcus woesii*, and its gene has been cloned and sequenced (74). It is a quite thermostable protein, with a half-life at 100°C of 44 min. Like *P. furiosus* glutamate dehydrogenase, GAPDH utilizes both NAD and NADP as electron carriers, showing that these nucleotides are present in these hyperthermophilic organisms. However, NADPH and NADH have half-lives of less than 5 min at 96°C in vitro (73), so some form of stabilization must occur for these cofactors to function in vivo at similar temperatures. One possibility is that they remain enzyme-associated rather than in free cytoplasmic "pools", although at present there is no evidence to support this.

Another question to be answered is why *P. furiosus* has chosen to use tungsten as a key element in its primary metabolic pathway, when the majority of biology utilizes the analogous element, molybdenum, as a active site redox component of a wide range of enzymes, e.g. nitrate reductase, formate dehydrogenase and xanthine oxidase (75). The reason for this is unknown, but among the extreme thermophiles this preference is not limited to *P. furiosus*. We have recently shown that the growth of the heterotrophs, *Thermococcus litoralis* and ES-4, is also stimulated by tungsten and tungsten-containing, O_2-sensitive enzymes analogous to *P. furiosus* AOR have been purified (Makund, S. and Adams, M. W. W., unpublished data). Interestingly, *T. litoralis* obtains energy for growth by peptide fermentation (31). The possibility therefore exists that these extremely organisms utilize this element in pathways for the metabolism of both peptides and carbohydrates.

Thermotoga maritima

Thermotoga maritima was isolated by Stetter and colleagues from geothermally-heated marine sediments in 1986 (13). As mentioned above, it is both the most thermophilic and the most ancient bacterium currently known. It is a rod-shaped organism with an unusual outer sheath that balloons over the ends. It is a strictly anaerobic heterotroph that obtains energy for growth by the fermentation of a range of both simple (glucose, ribose, xylose) and complex (starch, glycogen) carbohydrates. End-products of fermentation include H_2, CO_2, acetate and lactate. It grows up to 90°C with an optimum at 80°C. Like *P. furiosus*, it's growth is inhibited by the H_2 that it produces and this can be relieved either by sparging with an inert gas or by the addition of S°, which is reduced to H_2S. S° reduction does not appear to lead to any energy conservation. Like *P. furiosus*, *T. maritima* also grows well on complex media containing tryptone and yeast extract, and a peptide source is required for growth on carbohydrates since the organism does not seem to utilize ammonia or free amino acids as a N-source. We routinely obtain cell yields of *T. maritima* of approximately 700 g (wet weight) from a 400 liter culture (76).

Since *T. maritima* actively produces H_2 during growth, a primary aim of our research with this organism was to purify its hydrogenase and associated oxidoreductases. In our initial studies we found that its growth was also stimulated by the addition of tungsten to the growth medium, and this also caused the specific activity of the hydrogenase in cell-free extracts to increase by about 10-fold (76). However, the pure hydrogenase does not contain tungsten: it is a unique type of iron-only hydrogenase. To date, only a few hydrogenases lacking nickel are known, and they have all been found in strictly anaerobic mesophilic bacteria (77). These enzymes typically function to evolve H_2 in vivo, and use ferredoxin as an electron carrier. Their catalytic site is thought to comprise a novel type of iron-sulfur cluster, which gives rise to characteristic electron paramagnetic resonance (EPR) signals (77). In contrast to the mesophilic enzymes, *T. maritima* hydrogenase has very low H_2 evolution activity, does not use *T. maritima* ferredoxin as an electron carrier (see below), is inhibited by acetylene but not by nitrite, and does not exhibit the characteristic EPR signals of their catalytic sites, suggesting that it has a different site and/or mechanism for catalyzing H_2 production (76). The *T. maritima* enzyme is extremely O_2-sensitive, having a half-life in air of about 10 s, and is also intrinsically unstable. Like *P. furiosus* AOR, it has to be purified in buffer containing glycerol, DTT and sodium dithionite to prevent significant losses in activity. In spite of this, the pure enzyme is quite thermostable. It has a half-life at 90°C of about 1 hr, and the optimal temperature for catalysis is above 95°C (76).

A ferredoxin from *T. maritima* was also recently purified (Blamey, J. and Adams, M. W. W., unpublished data). This has a M_r value of about 7,000 and appears to contain a single [4Fe-4S] cluster. Although it will not donate electrons to the hydrogenase, it will accept them from the POR of this organism, which was also recently purified (Blamey, J. and Adams, M. W. W., unpublished data). The addition of tungsten to the growth medium also increases the specific activity of this enzyme in cell-free extracts by at least an order of magnitude. However, this also is not a W-containing enzyme. Like *P. furiosus* POR, *T. maritima* POR is a trimer of $M_r \sim 90,000$, but it does require added TPP for maximal catalytic activity, which occurs above 95°C. The inability of the ferredoxin to act as the sole electron carrier between POR and the hydrogenase suggests that *T. maritima* has additional electron carriers, perhaps other ferredoxins, that interact with the hydrogenase.

The strange effect of tungsten on the growth of *T. maritima* and the activities of hydrogenase and POR raises the question as to its physiological role. Our first thought was that *T. maritima* contains the W-dependent glycolytic pathway proposed for *P. furiosus* (Figure 3). However, this proved not to be the case, as no aldehyde or glucose oxidoreductase activities could be detected in cell-free extracts of *T. maritima.* The role or even necessity of tungsten in the growth of this organism is therefore a mystery at present.

In support of a more conventional glycolytic pathway in *T. maritima*, cell-free extracts contains exceedingly high amounts of GAPDH activity. This enzyme was recently purified (78) and was also sequenced (79). Like the analogous enzymes from mesophiles, it is a homotetramer of M_r 150,000. Its thermostability is comparable to that reported with some of the *P. furiosus* enzymes, in fact, it appears to be more thermostable than GAPDH from *P. furiosus.* The latter has a half-life at 100°C of 44 min (*74*), while that of *T. maritima* is about 2 hr, making it the most thermostable glycolytic enzyme known to date. The complete amino acid sequences of these two enzymes show in the range of 60% sequence identity with the mesophilic and moderately thermophilic GAPDHs that have been characterized. All of these enzymes have subunit M_r values close to 37,000, and both Jaenicke (*78*) and Hensel (*74*) and coworkers have performed sequence analyses to see if some insight could be obtained into the mechanisms of extreme thermostability. However, both concluded that the preferred amino acid exchanges in going to the more thermostable enzymes were not generally in accordance with previous studies, and that no generalizations could be made. In other words, there are no general "traffic rules" of thermal (or "cold") adaptation, and an understanding of the enhanced resistance of proteins to extreme temperatures can only come from local examinations of three-dimensional structures. Results from the first studies of this type using NMR spectroscopy and computer modelling are discussed in Chapters 3 and 11, respectively.

The only other metabolic enzyme that has been purified from *T. maritima* is lactate dehydrogenase, which catalyzes the NAD(H)-dependent interconversion of lactate and pyruvate (*80*). This is a homotetramer of M_r 144,000. Its molecular and catalytic properties are also very similar to the lactate dehydrogenases from mesophilic sources, except of course, in its thermostability. It has a half-life at 90°C of over 2 hr. Kinetic analyses indicated that under physiological conditions the enzyme catalyzes pyruvate reduction, which is in accordance with the fact that lactate is detected as a fermentation product in growing cultures (*13*). Although quantitative determinations have yet to be made, *T. maritima* appears to dispose of the excess reductant generated during fermentation both as H_2, via the hydrogenase, and as lactate, via lactate dehydrogenase. This is in spite of the fact that cell-extracts of *T. maritima* contain high POR activity. Presumably, POR and lactate dehydrogenase both utilize the pyruvate produced from glucose oxidation, generating acetyl CoA/H_2 and lactate/NAD, respectively. We have been unable to detect lactate dehydrogenase activity in cell-free extracts of *P. furiosus,* and lactate has not been reported as a fermentation product. These two organisms therefore appear to have different mechanisms for regenerating oxidized electron acceptors, in addition to different pathways of carbon flow during glucose oxidation.

Conclusions

In this chapter we have reviewed the organisms currently known that are able to grow near and above 100°C. The majority depend upon elemental sulfur, S°, for optimal growth. Most of the enzymes that have been purified from these organisms are involved in the primary metabolic pathways of two species, *Pyrococcus furiosus* and

Thermotoga maritima. Both obtain energy for growth by the oxidation of carbohydrates and the production of H_2. However, the pathways involved are very distinct. *P. furiosus* is thought to have a unique glycolytic pathway that is independent of NAD(P) and contains a novel aldehyde-oxidizing enzyme whose activity is dependent upon tungsten, an element seldom used in biological systems. *T. maritima* on the other hand has a more conventional pathway for oxidizing glucose. The majority of the enzymes involved are dramatically more thermostable than their counterparts in mesophilic organisms, particularly the *P. furiosus* enzymes. Amino acid sequence information is available for some of the enzymes but these offer little insight into the mechanisms of protein "hyperthermostability".

Acknowledgements

Research reported from the authors' laboratory was supported by grants from the Department of Energy (DOE FG09-88ER13901 to MWWA), the National Science Foundation (NSF DMB-8805255 to MWWA and BCS-9011583 to MWWA and RMK), and the Office of Naval Research (N00014-441204 to MWWA and N00014-89-J-1591 to RMK). The high temperature fermentation facility at the University of Georgia was funded by a grant from the National Institutes of Health (RR05577-01 to MWWA).

Literature Cited

1. Stetter, K. O. *Nature* **1982**, *300*, 258-260
2. Bryant, F. O.; Adams, M. W. W. *J. Biol. Chem.* **1989**, *264*, 5070-5079
3. Aono, S.; Bryant, F. O.; Adams, M. W. W. *J. Bacteriol.* **1989**, *171*, 3433-3439
4. Stetter, K. O. In *The Thermophiles;* Brock, T. D., Ed.; John Wiley, NY, 1986; pp39-74
5. Kelly, R. M.; Deming, J. W. *Biotech. Prog.* **1988**, *4*, 47-62
6. Adams, M. W. W. *FEMS Microbiol. Rev.* **1990**, *75*, 219-238
7. Stetter, K. O.; Fiala, G.; Huber, G.; Huber, R.; Segerer, G. *FEMS Microbiol. Rev.* **1990**, *75*, 117-124
8. Wiegel, J. K. W.; Ljungdahl, L. G. *CRC Crit. Rev. in Biotechnol.* **1986**, *3*, 39-107
9. Woese, C.R.; Fox, G. E. *Proc. Natl. Acad. Sci. USA* **1991**, *74*, 5088-5090
10. Woese, C. R.; Magrum, L. J. ; Fox, G. E. *J. Mole. Evol.* **1978**, *11*, 245-252
11. Woese, C. R.; Kandler, O.; Wheelis, M. L. *Proc. Natl. Acad. Sci. USA* **1990**, *87*, 4576-4579
12. Woese, C. R. *Microbiol. Rev.* **1987**, *51*, 221-271
13. Huber, R.; Langworthy, T. A.; König, H.; Thomm, M.; Woese, C. R.; Sleytr, U. B.; Stetter, K. O. *Arch. Microbiol.* **1986**, *144*, 324-333
14. Belkin, S.; Wirsen, C. O.; Jannasch, H. W. *Appl. Environ. Microbiol.* **1986**, *51*, 1180-1185

15. Jannasch, H. W.; Huber, R.; Belkin, S.; Stetter, K. O. *Arch. Microbiol.* **1988,** *150,* 103-104
16. Stetter, K. O. ; Zillig, W. In: *The Bacteria;* Woese, C. R. and Wolfe, R. S., Eds.; Academic Press, NY, 1985, Vol. VIII; pp 85-170,
17. Zillig, W.; Stetter, K. O.; Schäfer, W.; Janekovic, D.; Wunderl, S.; Holz, I.; Palm, P.; *Zbl. Bakt. Hyg., I Abt. Orig.* **1981,** *C 2,* 200-227
18. Fiala, G.; Stetter, K. O.; Jannasch, H. W; Langworthy, T. A.; Madon, J. *Syst. Appl. Microbiol.* **1986,** *8,* 106-113
19. Jannasch, H. W.; Wirsen, C. O.; Molyneaux, S. J.; Langworthy, T. A. *Appl. Environ. Microbiol.* **1988,** *54,* 1203-1209
20. Zillig, W.; Stetter, K. O.; Prangishvilli, D.; Schafer, W.; Wunderl, S.; Jankekovic, D.; Holz, J.; Palm, P. *Zbl. Bakt. Hyg., I Abt. Orig.* **1982,** *C3,* 304-317
21. Bonch-Osmolovskaya, E. A.; Slesarev, A. I.; Miroshnichenko, M. L.; Svetlichnaya, T. P.; Alekseev, V. A. *Mikrobiologiya* **1988,** *57,* 78-85
22. Zillig, W.; Gierl, G.; Schreiber, G.; Wunderl, S.; Janekovic, D.; Stetter, K. O.; Klenk, H. P. *Syst. Appl. Microbiol.* **1983,** *4,* 79-87
23. Huber, R.; Kristjansson, J. K.; Stetter, K. O. *Arch. Microbiol.* **1987,** *149,* 95-101
24. Segerer, A.; Neuner, A.; Kristjansson, J. K.; Stetter, K. O. *Intl. J. Syst. Bacteriol.* **1986,** *36,* 559-564
25. Zillig, W.; Yeats, S.; Holz, I.; Böck, A.; Rettenberger, M.; Gropp, F.; Simon, G. *Syst. Appl. Microbiol.* **1986,** *8,* 197-203
26. Stetter, K. O.; König, H.; Stackebrandt, E. *Syst. Appl. Microbiol.* **1983,** *4,* 535-551
27. Pley, U.; Schipka, J.; Gambacorta, A.; Jannasch, H. W.; Fricke, H.; Rachel, R.; Stetter, K. O. *Syst. Appl. Microbiol.* **1991,** *14,* 245-253
28. Fiala, G.; Stetter, K. O. *Arch. Microbiol.* **1986,** *145,* 56-61
29. Zillig, W.; Holz, I.; Klenk, H.-P.; Trent, J.; Wunderl, S.; Janekovic, D.; Imsel, E.; Hass, B. *Syst. Appl. Microbiol.* **1987,** *9,* 62-70
30. Zillig, W.; Holtz, I.; Janekovic, D.; Schafer, W.; Reiter, W. D. *Syst. Appl. Microbiol.* **1983,** *4,* 88-94
31. Neuner, A.; Jannasch, H.; Belkin, S.; Stetter, K. O. *Arch. Microbiol.* **1990,** *153,* 205-207
32. Zillig, W.; Holz, I. ; Wunderl, S. *Int. J. Syst. Bacteriol.* **1991,** *41,* 169-170
33. Zillig, W.; Holz, I.; Janekovic, D.; Klenk, H.-P.; Imsel, E.; Trent, J.; Wunderl, S.; Forjaz, V.H.; Coutinho, R.; Ferreira, T. *J. Bacteriol.* **1991,** *172,* 3939-3965
34. Pledger, R. J.; Baross, J. A. *Syst. Appl. Micrrobiol.* **1989,** *12,* 249-256
35. Pledger, R. J.; Baross, J. A. *J. Gen. Microbiol.* **1990,** *137,* 203-213
36. Stetter, K. O. *Syst. Appl. Microbiol.* **1988,** *10,* 172-173
37. Zellner, G.; Stackebrandt, E.; Kneifel, H.; Messner, P.; Sleytr, U. B.; de Macario, E. C.; Zabel, H.-P.; Stetter, K. O.; Winter, J. *Syst. Appl. Microbiol.* **1989,** *11,* 151-160
38. Stetter, K. O., Lauerer, G., Thomm, M. and Neuner, A. *Science* **1987,** *236,* 822-824

39. Achenbach-Richter, L.; Stetter, K. O.; Woese, C. R. *Nature* **1987**, *327*, 348-349
40. Jones, W. J.; Leigh, J. A.; Moyer, F.; Woese, C. R.; Wolfe, R. S. *Arch. Microbiol.* **1983**, *136*, 254-261
41. Stetter, K. O.; Thomm, M.; Winter, J.; Wildegruber, G.; Huber, H.; Zillig, W.; Janekovic, D.; König, H.; Palm, P.; Wunderl, S. *Zbl. Bakt. Hyg., I Abt. Orig.* **1981**, *C 2*, 166-178
42. Lauerer, G.; Kristjansson, J. K.; Langworthy, T. A.; König, H.; Stetter, K. O. *Syst. Appl. Microbiol.* **1986**, *8*, 100-105
43. Kurr, M.; Huber, R.; König, H.; Jannsch, H. W.; Fricke, H.; Trincone, A.; Kristjansson, J. K.; Stetter, K. O. *Arch. Microbiol.* **1991**, *156*, 239-247
44. Huber, R.; Stoffers, P.; Cheminee, J. L.; Richnow, H. H.; Stetter, K. O. *Nature* **1990**, *345*, 179-181
45. Brown, S. H.; Kelly, R. M. *Appl. Environ. Microbiol.* **1989**, *55*, 2086-2088
46. Weimer, P. J. In *The Thermophiles;* Brock, T. D., Ed.; John Wiley, NY, 1986; pp 217-257
47. Deming, J. W. *Microb. Ecol.* **1986**, *12*, 111-119
48. Stetter, K. O.; Zillig, W. In: *The Bacteria;* Woese, C. R.; Wolfe, R. S., Eds.; Academic Press, NY, 1985, Vol. VIII; pp 85-170,
49. Kjems, J.; Leffers, H.; Olsen, T.; Garrett, R. A. *J. Biol. Chem.* **1989**, *264*, 17834-17837
50. *Data for Biochemical Research;* Dawson, R. M. C.; Elliott, D. C.; Elliott, W. H.; Jones, K. M., Eds.; Clarendon Press, Oxford, 1986, 3rd Edn.
51. Masazumi, M.; Becktel, W. J.; Matthews, B. W. *Nature* **1988**, *334*, 406-410
52. Menedez-Arias, L.; Argos, P. *J. Mole. Biol.* **1989**, *206*, 397-406
53. Privalov, P. L.; Gill, S. J. *Advs. Prot. Chem.* **1989**, *39*, 191-254
54. Ahern, T. J.; Klibanov, A. M. *Science* **1985**, *228*, 1280-1284
55. Jaenicke, R. *Ann. Rev. Biophys. Bioeng.* **1981**, *10*, 1-67
56. Koch, R.; Zablowski, P.; Spreinat, A.; Antranikian, G. *FEMS Microbiol. Lett.* **1990**, *71*, 21-26
57. Koch, R.; Spreinat, A.; Lemke, K.; Antranikian, G. *Arch. Microbiol.* **1991**, *155*, 572-578
58. Constantino, H. R.; Brown, S. H.; Kelly, R. M. *J. Bacteriol.* **1990**, *172*, 3654-3660
59. Makund, S.; Adams, M. W. W. *J. Biol. Chem.* **1990**, *265*, 11508-11516
60. Yamamoto, I.; Saiki, T.; Liu, S.-M.; Ljungdahl, L.G. *J. Biol. Chem.* **1983**, *258*, 1826-1832.
61. Makund, S.; Adams, M. W. W. *J. Biol. Chem.* **1991**, *266*, 14208-14216
62. Brushi, M.; Guerlesquin, F. *FEMS Microbiol. Rev.* **1988**, *54*, 155-176
63. Park, J.-B.; Fan, C.; Hoffman, B. M.; Adams, M. W. W. *J. Biol. Chem.* **1991**, *266*, 19351-19356
64. Conover, R. C.; Kowal, A. T.; Fu, W.; Park, J.-B.; Aono, S.; Adams, M. W. W.; Johnson, M. K. *J. Biol. Chem.* **1990**, *265*, 8533-8541

65. Conover, R. C.; Park, J.-B.; Adams, M. W. W.; Johnson, M. K. *J. Am. Chem. Soc.* **1991,** *113,* 2799-2800
66. Conover, R. C.; Park, J.-B.; Adams, M. W. W.; Johnson, M. K. *J. Am. Chem. Soc.* **1990,** *112,* 4562-4564
67. Fauque, G.; Peck, H. D.; Jr.; Moura, J. J. G.; Huynh, B. H.; Berlier, Y.; DerVartanian, D. V.; Teixeira, M.; Przybyla, A.; Lespinat, P. A.; Moura, I.; LeGall, J. *FEMS Microbiol. Rev.*1989, *54,* 299-344
68. Danson, M. J. *Advs. Microbiol. Physiol.* **1989,** *29,* 165-231
69. Wahl, R. C.; Orme-Johnson, W. H. *J. Biol. Chem.* **1987,** *262,* 10489-10496
70. Schäfer, T.; Schönheit, P. *Arch. Microbiol.* **1991,** *155,* 366-377
71. Blumentals, I. L.; Robinson, A. S.; Kelly, R. M. *Appl. Environ. Microbiol.* **1990,** *56,* 1992-1998
72. Eggen, R.; Geerling, A.; Watts, J.; de Vos, W. M. *FEMS Microbiol. Lett.* **1990,** *71,* 17-20
73. Robb, F. T.; Park, J.-B.; Adams, M. W. W. *Biochim. Biopohys. Acta* **1992,** in the press
74. Zwickl, P.; Fabry, S.; Bogedain, C.; Haas, A.; Hensel, R. *J. Bacteriol.* **1990,** *172,* 4329-4338
75. *Molybdenum Enzymes;* Spiro, T. G., Ed.; Metals in Biology; John Wiley, NY, 1985; Vol. 7
76. Juszczak, A.; Aono, S.; Adams, M. W. W. *J. Biol. Chem.* **1991,** *266,* 13834-13841
77. Adams, M.W.W. *Biochim. Biophys Acta* **1990,** *1020,* 115-145
78. Wrba, A.; Schweiger, A.; Schultes, V.; Jaenicke, R.; Zavodszky, P. *Biochemistry* **1990,** *29,* 7584-7592
79. Schultes, V.; Deutzmann, R.; Jaenicke, R. *Eur. J. Biochem.* **1990,** *192,* 25-31
80. Wrba, A.; Jaenicke, R.; Huber, R.; Stetter, K. O. *Eur. J. Biochem.* **1990,** *188,* 195-201

RECEIVED March 3, 1992

Chapter 3

Characterization of Enzymes from High-Temperature Bacteria

Robert M. Kelly[1], S. H. Brown[2], I. I. Blumentals[3], and Michael W. W. Adams[4]

[1]Department of Chemical Engineering, North Carolina State University, Raleigh, NC 27695-7905
[2]Center for Marine Biotechnology, University of Maryland, Baltimore, MD 21202
[3]Department of Chemical Engineering, The Johns Hopkins University, Baltimore, MD 21218
[4]Department of Biochemistry and Center for Metalloenzyme Studies, University of Georgia, Athens, GA 30602

The purification and characterization of enzymes from bacteria that grow at extremely high temperatures presents numerous challenges. In addition to selecting and cultivating appropriate microorganisms to study, there are many other distinguishing features related to the study of "hyperthermophilic" enzymes. These are discussed and illustrated through case studies involving an α-glucosidase and a rubredoxin purified from the hyperthermophile, *Pyrococcus furiosus*, an organism that grows optimally at 100°C.

With the discovery of microorganisms inhabiting niches associated with geothermal activity, it has become clear that the upper temperature for life is at least 110°C (*1*). As such, it is also apparent that the molecular species responsible for essential life processes in these bacteria, such as proteins, nucleic acids, lipids, vitamins, ATP and coenzymes, must be either intrinsically stable or sufficiently stabilized to enable a cell to not only survive but to proliferate at extreme temperatures. Clearly, insights into the factors responsible for the structure and function of biomolecules at elevated temperatures will provide a basis for further study of high temperature microbial physiology, in addition to creating biotechnological opportunities. Of interest here is an understanding of biocatalysis at elevated temperatures.

There are many experimental and theoretical challenges that present themselves in the characterizing thermodynamic and kinetic properties of extremely thermostable enzymes. Working with novel biological systems creates problems, and this is especially true if these systems must be manipulated at high temperatures. However, such

0097-6156/92/0498-0023$06.00/0

problems must be overcome if meaningful extension is to occur of present knowledge concerning the influence of temperature on protein structure and function.

In addition to their physiological significance, the underlying factors that contribute to biocatalysis at extremely high temperatures are intriguing. So far it is not clear whether these factors can be elucidated using existing biochemical and biophysical dogma or if "new rules" must be invoked. Because so little is known about enzymes from bacteria growing at temperatures near and above 100°C, it is difficult to draw any generalizations about the basis for thermostability and activity. However, it is possible to provide some perspective on approaches that can be used for their study and to discuss some characteristics that have, thus far, been identified.

High Temperature Enzymology

In the study of any particular enzyme from a microorganism, some framework should be established to put the information gained in perspective. The framework can be based on any of several criteria; for example, physiological function, structural or catalytic characteristics, particular coenzyme or cofactor requirements, regulatory aspects or evolutionary considerations. One of the most attractive approaches to addressing enzyme structure and function at elevated temperatures is to choose an enzyme which has many well-characterized counterparts from less thermophilic sources. If many prevailing hypotheses are correct, it may be that a relatively small number of changes in amino acid sequence can impart significant thermostability to a particular enzyme. This might be sorted out if a series of homologous enzymes, with increasing thermostability, could be examined. However, predicting how these changes influence secondary and tertiary structure is beyond present capabilities. Site-directed mutagenesis, more often than not, results in an enzyme that is less active and less stable.

Nonetheless, the study of enzymes from high temperature bacteria could help to unravel some of the subtleties of protein stability. Immunological cross-reactivity between enzymes of similar function suggest significant structural similarities for enzymes with vastly different optimal temperatures (2). In addition, catalytic centers are apparently conserved over large temperature ranges. However, significant headway in understanding the factors underlying extreme thermostability has not yet been realized.

Choice of Microorganism

Currently, one of the central issues in choosing an enzyme to study from a particular high temperature bacterium is the ease of obtaining sufficient amounts of biomass from which the enzyme can be purified. Although genetic engineering offers a potential solution to this problem, the successful expression of hyperthermophilic proteins in more suitable hosts is still problematic. This is not to say that it has not been accomplished (e.g., 3), but more classical approaches to obtaining purified enzyme are mostly used at present. In any event, a certain amount of purified material is usually necessary to initiate cloning efforts.

One of the general truths that has emerged in the study of high temperature bacteria is that biomass yields are low relative to less thermophilic bacteria. Table I lists maximum cell densities and approximate biomass yields for several extremely thermophilic (optimum growth temperature between 75-95°C) and hyperthermophilic (optimum growth temperature above 95°C) bacteria. The low biomass yields are exacerbated by the sometimes unpredictable growth physiology of a particular bacterium. Under some conditions, no significant stationary phase may occur and extensive lysis will take place at the end of exponential phase (4). The reasons for this behavior are not clear. No phages have yet been identified in association with a high temperature bacterium, although there is some evidence that these might exist (5). The potential for lysis, of course, presents difficulties in the recovery of cellular material and complicates recovery of intracellular and extracellular enzymes. In addition to cell lysis, biomass yields on a large-scale are often unpredictable. This may be in large part due to the engineering problems of maintaining appropriate redox potentials, the need to balance mixing intensity against the shear sensitivity of many high temperature strains, and the use of manual approaches for temperature control. For instance, most fermentation systems are not designed for control temperature between about 60°C and sterilization temperatures.

Another key issue to consider in choosing a high temperature bacterium for study is the requirement of sulfur for growth. With some exceptions (e.g., *Pyrococcus furiosus, Thermococcus litoralis, Thermotoga maritima*), a reducible sulfur compound must be included in the growth medium for high temperature bacteria. If sulfur is included in the medium, high levels of sulfide are produced during growth although the physiological importance of sulfur reduction remains unclear (6,7). With high levels of sulfide, conventional fermentation systems constructed of stainless steel are impractical for biomass generation, thus greatly limiting the list of candidate organisms as sources of thermostable proteins. The pioneering work in this field by Stetter and co-workers (8) led to the development of large-scale enamel-lined fermentors; however, these systems are constructed at considerable expense. Brown and Kelly (9) described the use of a simple, scaleable continuous culture device for generating intermediate amounts of biomass. However, while sterility in the bioreactor is not a problem, contamination of feed stocks, often rich in complex extracts, can present problems in long-term runs. Until expression in mesophilic hosts becomes reliable, small amounts of high temperature enzymes, purified from continuous culture runs, could be an attractive solution to the present dilemma. Reactor volumetric productivity in such systems is high and problems with corrosivity and exposure to toxic hydrogen sulfide can be minimized. In fact, this approach seems to be the best generally available way to lengthen the current list of potential organisms for study beyond those not obligately requiring sulfur for growth.

Another issue to be considered in choosing the microbial source for a particular enzyme is whether the enzyme is produced constitutively, inducibly or under regulatory control. This is clearly an important issue for any enzyme, mesophilic or thermophilic. However, if enzyme expression is maximal under limiting nutrient conditions, the overall yield for the enzyme may be extremely low. This was found to be the case in the production of proteolytic activities from *P. furiosus*. As is the case with amylolytic enzymes from this particular bacterium (Brown and Kelly, submitted), the optimization of enzyme expression through nutritional experiments may be more important than with mesophilic systems.

Choice of Enzyme

Thus far, the choice of enzyme to study from high temperature bacteria has been based on existing expertise with comparable enzymes from less thermophilic origins or driven by biotechnological potential. In regard to the latter, the isolation and wide-spread use of DNA polymerase from *Thermus aquaticus* in polymerization chain reaction (PCR) technology has fueled interest in the identification of other polymerases (see Chapter 13). However, most reports to date have focused on physiological issues or previous interests in less thermophilic counterparts.

One of the first reports of the characteristics of an enzyme from a hyperthermophile suggested that there are similarities between mesophilic and hyperthermophilic enzymes. Pihl et al. (2) showed that antibodies directed against one subunit of an uptake hydrogenase from a mesophile, *Bradyrhizobium japonicum*, recognized a similar subunit from the hydrogenase from the hyperthermophile, *Pyrodictium brockii*. While there were marked differences in the temperature optima of the two enzymes, the immunological cross-reactivity suggested some structural relationship. Adams and co-workers, building on expertise on several metalloenzymes from less thermophilic bacteria, have purified and characterized several enzymes, including hydrogenases, rubredoxins and ferredoxins, from high temperature bacteria (10). In addition, they have purified and characterized a tungsten protein (11) implicated in a novel modification of an Entner-Doudoroff pathway apparently important in the physiology of several high temperature bacteria (Schicho et al., submitted: see Chapter 2). Another enzyme central to the EMP pathway, glyceraldehyde phosphate dehydrogenase, has been isolated from *Pyrococcus woesii*, cloned and expressed in *Escherichia coli* (3). The features of this enzyme, including its gene sequence, were compared to several mesophilic counterparts. Schafer and Schoneit (12) have also examined several enzymes implicated in pyruvate metabolism in *Pyrococcus furiosus*.

There has also been some focus on hydrolytic enzymes from hyperthermophilic and extremely thermophilic bacteria. Bragger et al. (13) screened a number of high temperature bacteria for proteolytic, amylolytic and cellulolytic activities and found these to be relatively widespread. High levels of proteolytic activity have been identified in *P. furiosus* (14-16) which might, in part, be related to the multicatalytic proteinase complex identified in eukaryotic cells (Snowden et al., in press). Several amylolytic activities have also been found in *P. furiosus* (*17*) including α-glucosidase, pullulanase and amylase. These activities have been found in many high temperature heterotrophs. Such hydrolases are, of course, potentially important as commercial enzymes.

Purification Protocols

Assuming that sufficient biomass can be generated containing an enzyme or enzymes of interest, purification protocols must be developed. Thus far, existing methodologies can be adapted and have been used in most cases reported to date. Affinity techniques have also been employed in some final purification steps (Schuliger et al., submitted; Brown et al., submitted; *2*) suggesting availability of the active site at low (ambient) temperatures. In general, because of the high stability of these enzymes and the very low proteolytic activity at ambient temperatures in cell-free extracts, purifications can be carried out at room temperature. However, some of the enzymes appear to be unstable at ambient temperature and stabilizing agents such as glycerol must be added during their purification (see Chapter 2). It has not yet been established as yet whether any enzyme from a high temperature bacterium is adversely affected by cold (4°C) storage although cold denaturation may be problematic in some instances.

There are, however, some factors that must be taken into account when purifying enzymes from high temperature bacteria. In some cases, their proteins have been shown to contain a high percentage of hydrophobic amino acids (Brown and Kelly, submitted), and this appears to lead to aggregation under some conditions. For example, purification of proteolytic species from *P. furiosus* has been problematic because of their aggregation and also their apparent association with other proteins. In one case, extracts of *P. furiosus* were boiled in denaturing buffer for periods of 24 hours or longer resulting in the purification of a 66 kDa serine protease. A combination of proteolysis and thermal and chemical denaturation was apparently implicated. Without this treatment, all proteolytic activity was found to aggregate and drop out in the first purification step. There have been other instances in which two or more seemingly unrelated proteins will co-purify, necessitating the use of alternative purification steps. Hydrophobic interaction chromatography has been especially useful in this regard.

Another problem that is encountered is the difficulty in assessing the effectiveness of a particular purification step. For example, the conditions for SDS-polyacrylamide gel electrophoresis (SDS-PAGE) must be modified so that the protein(s) of interest from high temperature bacteria is truly denatured. Schuliger et al. (submitted) showed that an amylopullulanase from a novel marine hyperthermophile, ES4, had to be incubated for 10 minutes at 122°C before the enzyme was denatured. When conventional denaturing buffer treatments are used prior to SDS-PAGE, there can be problems with reproducibility, especially if preparatory steps were not consistent from one sample to another. As a result, the apparent molecular weight of a given protein can vary thus making the efficacy of a particular purification step difficult to determine. In addition, as discussed below, care must be also taken to ensure that activity assays are representative and accurate. If not, purification protocols will be difficult to develop and evaluate.

Enzymatic Assays at High Temperatures

Given the growth temperature of the microorganism from which they are to be isolated, enzymes from extreme thermophiles are usually active at significantly higher temperatures than their mesophilic counterparts. Thus, appropriate assays must be developed to assess activity at high temperatures. This can present significant problems. For example, many substrates that could be used may be very unstable at higher temperatures leading at best to significant corrections relative to controls. At worst, alternative substrates may have to be developed. Enzymes requiring co-factors, e.g., NADH, are also difficult to study at high temperatures because of the thermolability of the coenzyme (see Chapter 6), despite their attractiveness from a the perspective of comparative enzymology. Another problem is that many high temperature bacteria will produce enzymes which are optimally active well above 100°C (18). Enzyme assays at such temperatures will require the use of pressure to prevent boiling and achieve meaningful results. From a practical perspective, the handling of samples at high temperatures can significantly slow the progress of research efforts.

Case Studies - High Temperature Enzymes

To illustrate some of the points made in the discussion above, we will use preliminary results from two proteins isolated from P. furiosus.

Case I - α-glucosidase. We have shown that P. furiosus produces a range of proteolytic (14; Snowden et al., submitted) and amylolytic species (17; Brown and Kelly, submitted), all of which possess high thermostability. Initial efforts in the study of these hydrolases focused on the α-glucosidase from P. furiosus (18,19). This enzyme was chosen for several reasons. First, the intracellular breakdown of maltose and other small saccharides to glucose, which are the reactions catalyzed by this enzyme, is an important step in the utilization of saccharides by P. furiosus and related bacteria. Second, a p-nitrophenol (PNP) substrate-linked assay was available which is effective at temperatures well in excess of 100°C and no cofactor requirement. Third, this enzyme does not appear to be sensitive to aerobic conditions despite coming from a strictly anaerobic bacterium. And fourth, α-glucosidase has mesophilic counterparts that have been characterized to some extent. Our initial efforts with the enzyme from P. furiosus showed it to be monomeric with an approximate molecular weight of 140,000 kDa. It can be readily purified from cell-free extracts and apparently exists as an intracellular enzyme (Figure 1).

Table I. Biomass Yields of Different Extremely Thermophilic Bacteria

Organism	Cell Density (cells/ml)	Cell Yield [g (wet wt.)/l]
Pyrococcus furiosus	2.0 - 5.0 E+08	0.75-1.5
Pyrococcus woesei	3.0 E+08	1.00
Pyrodictium brockii	2.0 E+07	0.10
Thermococcus litoralis	5.0 E+08	1.00
Metallosphaera sedula	5.0 E+07	0.10
Thermotoga maritima	2.0 E+08	0.75
Staphylothermus marinus	3.0 E+08	0.30
Archaeglobus fulgidus	2.0 E+08	0.50
Sulfolobus acidocaldarius	8.0 E+08	1.00

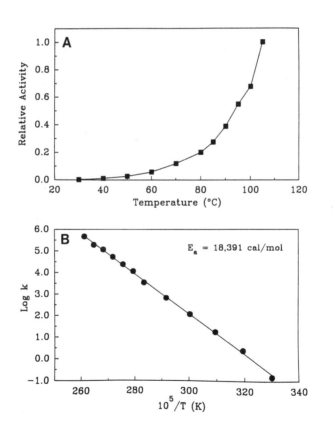

Figure 1. A. Effect of temperature on the α-glucosidase activity of *P. furiosus*. Activities were determined using α-PNPG in 50 mM phosphate buffer, pH 6.0. B. Arrhenius plot of the data presented in panel A.

α-glucosidase was purified from *P. furiosus* cells grown in both the presence and absence of elemental sulfur. A sample purification protocol is shown in Table II. Papers by Costantino et al. (*18*) and Brown (*19*) provided some preliminary characterization of the enzyme. It has an optimal temperature above 100°C, significantly higher than previous enzymes of this type (see Table III). Enzyme kinetics follow a Michaelis-Menten formulation which allowed us to determine apparent V_{max} and K_m values over a wide temperature range (see Table IV). From this data, an activation energy (E_a) could be determined which surprisingly was constant over 8 decades of temperature. This was not the case for two other glucosidases isolated from moderate thermophiles as these exhibited biphasic activation energy plots (*20,21*). The significance of this result is not completely clear, although it suggests that the active site is formed at mesophilic temperatures and remains functional to very high temperatures. Another implication is that the α-glucosidase from *P. furiosus* is a rigid molecule, maintaining its rigidity over the temperature range studied. This could be a factor in the extreme thermostability of this enzyme. Analysis of the amino acid sequence derived from sequence of the GAP-dehydrogenase from *P. woesii* as compared to mesophilic counterparts leads to a similar conclusion concerning the role of rigidity in thermostability (*3*). The *P. furiosus* α-glucosidase also exhibits considerable stability when pre-incubated at 98°C for 30 minutes in the presence of several common protein denaturants (see Table V). Maintenance of activity in the presence of high levels of DTT most likely indicates that disulfide linkages are not important contributors to the stability of this protein. In addition, SDS does not completely denature the enzyme, and so, as discussed above, caution must be used in the interpretation of SDS-PAGE results.

Having completed an initial characterization of the α-glucosidase, advanced biochemical methods were sought to further study the nature of its thermostability. Among the techniques available, differential scanning microcalorimetry (DSC) was selected as it potentially could yield detailed thermodynamic information. In particular, we hoped that by utilizing DSC we could determine the difference in heat capacity, ΔC_p, between the folded and unfolded protein. This quantity results from the exposure of buried nonpolar hydrophobic amino acids to the polar aqueous environment, and is, therefore, a direct measure of the hydrophobic contribution to the protein's stability. Once ΔC_p has been measured, the stability of the protein, calculated as the difference in free energy (ΔG) between the folded and unfolded forms, can be predicted over the temperature range of interest. For example, this prediction would allow us to assess whether the cold denaturation temperature was above 0°C and, therefore, accessible to experimental observation. However, this technique, even with its relatively high level of sensitivity, required significant amounts of protein for a comprehensive study. This presented a problem as purification yields for the α-glucosidase typically were on order of 10 mg of the purified protein from biomass generated in a 400 liter fermentation. Nonetheless, the study was

Table II. Purification of α-glucosidase from *P. furiosus*

Fraction	Volume (ml)	Activity (U/ml)	Protein (mg/ml)	Specific Activity (U/mg)	Fold Purification	% Yield
Crude extract	50.0	15.1	19.3	0.782	1.0	100.0
Ammonium sulfate	50.0	12.6	13.3	0.950	1.2	84.0
Anion exchange	108.0	2.83	0.508	5.58	7.2	41.0
Hydroxyapatite	36.0	5.78	0.270	21.4	27.0	28.0
Gel filtration	36.0	4.61	0.0772	59.9	77.0	22.0
Electrelution	4.1	15.6	0.0638	245.0	310.0	8.5

Table III. Comparison of Thermostable α-Glucosidases

Organism (T_{opt})	(T_{opt}) (°C)	M.W. (kD)	K_m (mM)	V_{max} (U/mg)	E_a (kcal/mol)
B.caldovelox [a] (60°C)	60	30	NA	140	NA
B.stearothermophilus[b] (60°C)	70	47	0.63	123	biphasic 17 (13-40 °C) 12 (40-70 °C)
B. thermoglucosidus[c] (60°C)	75	55	0.23	183	biphasic 28 (30-50 °C) 5.5 (65-75 °C)
T.thermophilus[d] (75°C)	80	67	0.4	70	NA
P.furiosus (100°C)	>110	140	0.11	310	18

[a]Ref. 43. [b]Ref. 44. [c]Ref. 45. [d]Ref. 46

Table IV. Michaelis-Menten parameters for *P. furiosus* α-glucosidase[a]

°C	Vmax (U/mg)	Km (mM)
60	9.04	0.018
70	19.14	0.019
80	40.75	0.026
90	84.44	0.039
98	139.84	0.047
110	310.00	0.110

[a]Determined using α-PNPG in 50 mM acetate buffer, pH 5.6.

Table V. Effect of various denaturing and chelating agents
on *P. furiosus* α-glucosidase

Reagent	Remaining Activity after 30 min at 98°C (%)
100 mM DTT	83
1.0 M Urea	84
1.0 M Guanidine hydrochloride	<1.0
1 % SDS	22

initiated even with the prospect of needing to process thousands of liters of culture media.

Preliminary results obtained from the initial calorimetry study are shown in Figure 2. The instrument used for these studies had as an upper temperature limit approximately 105°C, but this was not high enough to bring about thermal denaturation. Thus, a chaotropic agent, guanidine hydrochloride (GuHCl) was used to facilitate the unfolding process (22). Figure 3 shows that protein unfolding occurred at lower temperatures as the amount of GuHCl was increased. Unfortunately, these results indicated that a significant amount of aggregation (as evidenced by the large exotherm after the peak in the endotherm and the failure of the response curve to return to baseline) resulted as the α-glucosidase unfolded through thermal denaturation in the presence of GuHCl, presumably because of the exposure of a large hydrophobic core. This presents problems in gleaning thermodynamic information from this data in that the expected endotherm resulting from the unfolding of the protein must be separated from the exothermic contribution arising from protein aggregation. The baseline on the high temperature side is difficult to determine, which complicates data analysis.

In addition, the scan for the 1.5 M GuHCl case suggested that there could be more than one species in the sample. Further purification effort confirmed this; in addition to the α-glucosidase, a second protein of similar molecular weight had co-purified. The presence of the second protein (named P1) was most likely the result of the elimination of the electroelution step used by Costantino et al. (18) in the original purification protocol. This step is impractical for purifying multi-milligrams of the α-glucosidase necessary for microcalorimetry studies. However, the use of hydrophobic interaction chromatography has allowed us to separate α-glucosidase from the P1 protein and obtain both in homogeneous forms.

We then examined pure α-glucosidase, separated from the P1 protein, by microcalorimetry. However, as shown in Figure 4, significant aggregation of the protein occurred, even in the presence of GuHCl. This rendered the quantitative treatment of the data infeasible for the reasons discussed above. Fortunately, further efforts with the co-purified protein, P1 (to this point, its function has not been identified), resulted in meaningful calorimetry experiments as shown in Figure 5. This protein is a homotetramer with a molecular weight of approximately 140 kDa. Efforts to identify the function of this protein, to clone and determine its gene sequence, and conduct detailed thermodynamic analysis are underway (Straume et al., in preparation).

In hindsight, the choice of the α-glucosidase as a model hyperthermophilic protein was both good and bad. Because of its physiological significance, ease of assay at elevated temperatures, and numerous counterparts, there are advantages to its choice. However, because it is a very large, monomeric enzyme with a large hydrophobic core, techniques for its characterization involving protein unfolding may not be useful. Also, the difficulties in obtaining significant

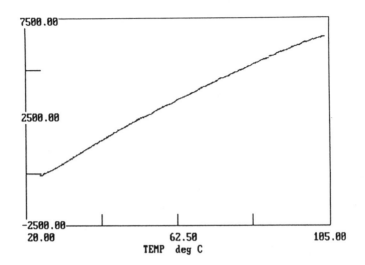

Figure 2. Initial DSC scan of the α-glucosidase from *P. furiosus*. The enzyme concentration was 0.8 mg/ml in 50 mM phosphate buffer, pH 6.0. The ordinate is heat capacity expressed in units of mcal·g⁻¹· K⁻¹.

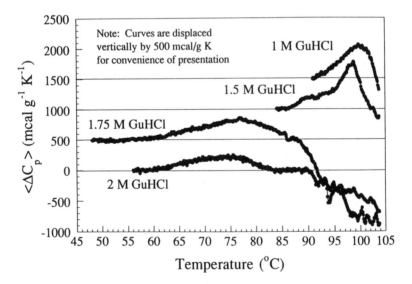

Figure 3. DSC scans of *P. furiosus* α-glucosidase (1.0 mg/ml) at varying concentrations of guanidine hydrochloride in 50 mM phosphate buffer, pH 6.0. Scans are baseline-subtracted and have been displaced vertically by 500 mcal.g⁻¹.K⁻¹ for clarity. The enzyme preparation used also contained the P1 protein (see text for details).

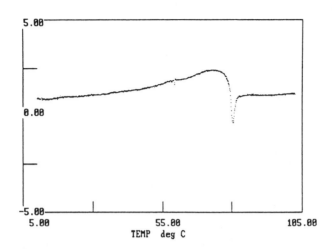

Figure 4 A DSC scan of pure *P. furiosus* α-glucosidase (4.0 mg/ml) in 50 mM phosphate buffer, pH 6.0, containing 2.5 M guanidine hydrochloride. The ordinate is heat capacity in units of mcal·K^{-1}.

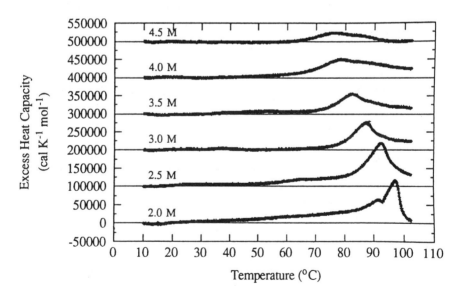

Figure 5. DSC scans of pure P1 protein (4.0 mg/ml) from *P. furiosus* at varying guanidine hydrochloride (GuHCl) concentrations in 50 mM phosphate buffer, pH 6.0. Scans have been displaced vertically by 500 mcal·g^{-1}·K^{-1} for clarity and are baseline-subtracted.

amounts of the enzyme preclude extensive experiments to determine conditions minimizing aggregation problems. Thus, other enzymes from high temperature bacteria, such as the unidentified P1 protein that co-purified with the α-glucosidase, will be better candidates for certain types of studies.

Case 2 - Rubredoxin. Rubredoxins are a class of bacterial electron-transfer proteins that contain a single iron atom coordinated by the sulfur atoms of four cysteinyl residues (23). They have been purified and characterized from several types of strictly anaerobic organisms, and although its physiological role in these organisms is still unclear, so far they constitute an homologous group of proteins. All are monomers of molecular weight approximately 6,000, and have midpoint potentials near 0 mV. Complete amino acid sequences of eleven rubredoxins have been reported (24-36) and four crystal structures are also known (37-40). As the simplest known class of redox protein, the study of rubredoxins has the potential to provide insight into the minimum structural requirements for this protein and the residues that influence electron transfer and the redox properties of the iron site, in addition to more generic information on protein stability and phylogenetic relationships. The only structural information available on rubredoxins so far comes from those purified from bacteria, all but one of which are from mesophilic species. The exception is the protein from *Clostridium thermosaccharolyticum* (27), which grows optimally at 55°C.

Our recent characterization of the rubredoxin from *P. furiosus* was therefore of some significance (41). It is the first to be purified from an archaebacterium, and also the first rubredoxin from a hyperthermophilic organism. In many ways the properties of the hyperthermophilic protein are very similar to those of the rubredoxins that have been previously purified from mesophilic eubacteria. These include molecular size (M_r = 5,400), Fe content (1 atom/mole), midpoint potential (0 mV at 23°C, pH 8) and EPR and UV-visible absorption properties. However, the *P. furiosus* protein is extremely thermostable, being unaffected by 24 hr at 95°C. This compares with the rubredoxin from *Clostridium pasteurianum*, which is rapidly denatured at 80°C (24), from *Desulfovibrio gigas*, which loses 50% of its visible absorption (t_{50}%) after approximately 2 hr at 80°C (42), and from *Thermodesulfobacterium commune*, which has a t_{50}% value of about 6 hr at 80°C (42). The question is, are there any indications of the mechanisms responsible for the enhanced thermostability of the *P. furiosus* protein from its amino acid sequence ?

The sequence of *P. furiosus* rubredoxin is aligned with those of the rubredoxins from ten mesophilic eubacteria and one moderately thermophilic eubacterium (*C. thermosaccharolyticum*) in Figure 6. Some 15 conserved residues were previously found in a comparison of ten rubredoxin sequences (27), and these were also present in the subsequently published sequence of rubredoxin from *Desulfovibrio vulgaris* strain Miyazaki (31). Of these, 14 are also conserved in *P. furiosus* rubredoxin, the exception being the N-terminal methionine residue. The consensus residues are the four cysteinyl residues that

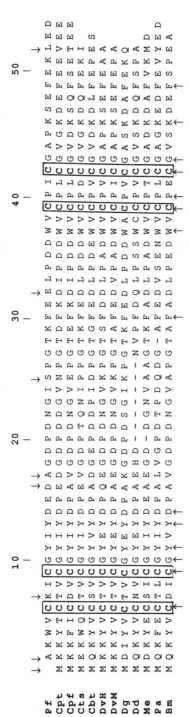

Figure 6. Comparison of rubredoxin amino acid sequences from anaerobic bacteria. The abbreviations are: Pf, *P. furiosus*; Cpt, *Clostridium pasteurianum*; Cpf, *C. perfringens*; Cts, *C. thermosaccharolyticum*; Cbt, *Chlorobium thiosulfatophilum*; DvH, *Desulfovibrio vulgaris* strain Hildenborough; DvM, *Desulfovibrio vulgaris* strain Miyazaki; Dg, *D. gigas*; Dd, *D. desulfuricans*; Me, *Megasphaera elsdenii*; Pa, *Peptococcus aerogenes*; Bm, *Butyribacterium methylotrophicum* (14). Conserved cysteines are given in bold. Downward arrows indicate residues (or lack of) unique to Pf rubredoxin and upward arrows indicate residues conserved in all known rubredoxins. The four cysteinyl residues that bind the Fe atom are boxed. (Adapted from reference 41.)

bind the Fe atom together with an adjacent prolyl residue, five aromatic residues, which from crystal studies constitute a hydrophobic core (37-40), two glycyl residues, and one lysyl and one aspartyl residue. Other than these, there are several positions that are occupied by only one residue in most of the sequences, but the exceptions show that a particular type of residue cannot be essential for the structure or function of rubredoxins in general. For example, all contain a prolyl residue at position 15, except the P. furiosus and M. elsdenii proteins, which contain a glutamyl group.

The overall sequence homology (identity) of P. furiosus rubredoxin with the other rubredoxins (Figure 6) ranges from 42% to 67%, which is similar to the homologies found between the other rubredoxins (41% to 90%). Seven residues are found in P. furiosus rubredoxin but in none of the others (excluding the absence of the N-terminal methionine). A previous study of the rubredoxin from the moderate thermophile, C. thermosaccharolyticum identified four unique residues that might render this protein more stable compared with those of mesophilic species. These included a tryptophan residue at position 4, which is also present in the P. furiosus protein. However, the other three unique residues (thr, pro and asp at positions 21, 25 and 41, respectively) are not found in the hyperthermophilic protein (asp, ser and ile at positions 21, 25 and 41, respectively). Previous studies comparing mesophilic and thermophilic proteins together with one from a hyperthermophilic organism (glyceraldehyde-3-phosphate dehydrogenase from Pyrococcus woesii) have shown a preference for alanine and discrimination against serine and glycine in the thermophilic proteins, and a striking preference for phenylalanine, and discrimination against aspartate, methionine and cysteine in the hyperthermophilic one (3). However, none of these preferences appear to be a general feature of 'hyperthermostability' since none are evident when one compares the hyperthermophilic rubredoxin with the others listed in Figure 6. Indeed, systematic comparisons show no obvious preference for any amino acid in the latter protein, with the exception of isoleucine (4 in P. furiosus rubredoxin, 3 in the C. thermosaccharolyticum protein, and 2 or less in the others).

It is, therefore, obvious that meaningful insight into mechanisms of enhanced thermostability in general, and of hyperthermostability in particular, can only come from the analysis of three-dimensional structures. We have embarked upon collaborative studies to approach this problem using three complementary techniques. The first is by protein crystallography. Crystals of P. furiosus rubredoxin suitable for X-ray diffraction studies have been recently obtained, and data collection and analyses are now in progress (Rees, D., unpublished data). The second is using molecular modelling and dynamics, and current progress in this area is summarized in Chapter 11. The third approach is through the use of high resolution two-dimensional NMR spectroscopy (41). Unfortunately, 2D NMR analysis of the oxidized and reduced forms of P. furiosus rubredoxin showed substantial line broadening due the paramagnetic iron atom in these two states, and this precluded detailed 3D studies. However, high quality spectra have been obtained from the diamagnetic zinc-substituted protein. Due to the complexity of the spectra and of the analysis required to interpret them (see *41*), only a summary of the conclusions will be given here.

P. furiosus rubredoxin was shown by NMR to adopt a 3D conformation very similar to that of the prototypical rubredoxin, that of *Clostridium pasteurianum*. Both proteins contain a triple-stranded, antiparallel β-sheet with several tight turns and a hydrophobic core. The β-sheet domain is more extensive in the *P. furiosus* protein than it is in the mesophilic one and comprises about 30% of the primary sequence. This domain also differs from that in *C. pasteurianum* rubredoxin in that it begins at the N-terminal residue. This presumably gives added rigidity to the protein. Perhaps the most significant result from the NMR analysis is that potential stabilizing electrostatic interactions involving the charged residues ala(2), glu(15) and glu(53) were identified. These interactions are not possible in any of the other rubredoxins listed in Fig. 6 as they all have an "extra" methionine at their N-termini. Moreover, all but one have a prolyl residue rather than a glutamyl group at position 15. The significance of these interactions is not known but they could well lock the N-terminal into the main structure of the protein and prevent it from "unzipping" at high temperatures.

Summary

As should be clear from the discussion presented here, approaches used for the study of high temperature enzymes will utilize existing experimental and theoretical methodologies as well as require novel developments. Fundamental insights concerning the basis for extreme thermostability are beginning to emerge but many obstacles must be overcome before significant progress can be made. As additional experimental data are obtained and analyzed in the appropriate theoretical framework, high temperature enzymes and proteins, compared and contrasted with their mesophilic counterparts, will likely expand our understanding of protein chemistry and biocatalysis.

Acknowledgements

The work described here was supported, in part, through grants from the National Science Foundation (BCS-9011583, BCS-8813608) and the Office of Naval Research (N00014-441204 to MWWA and N00014-89-J-1591 to RMK). The authors would like to thank Dr. Martin Straume of the Biocalorimetry Center, Johns Hopkins University, for assistance with the calorimetry experiments and helpful discussions.

Literature Cited

1. Stetter, K.O.; Fiala, G.; Huber, G.; Huber, R.; Segerer, G. *FEMS Microbiol. Rev.* **1990**, 75, 117-124.
2. Pihl, T.D.; Schicho, R.N.; Kelly, R.M.; Maier; R.J. *Proc. Nat. Acad. Sci. USA.* **1989**, *86*, 138-141.

3. Zwickl, P.; Fabry, S.; Bogedain, C.; Haas, A.; Hensel, R. *J. Bacteriol.* **1990**, *172*, 4329-4338.
4. Malik, B.; Su, W-.W.; Wald, H.L.; Blumentals, I.I.; Kelly, R.M. *Biotechnol. Bioeng.* **1989**, *34*, 1050-1057.
5. Zillig, W.; Holz, I.; Klenk, H-.P.; Trent, J.; Wundrel, S.; Janeko-vic,D.; Imsel, E.; Haas, B. *System. Appl. Microbiol.* **1987**, *9*, 62-70.
6. Blumentals, I.I.; Itoh, M.; Olson, G.J.; Kelly, R.M. *Appl. Environ. Microbiol.* **1990**, *56*, 1255-1262.
7. Schicho, R.N. Ph.D. Thesis, Johns Hopkins University, Baltimore, MD., **1992**.
8. Stetter, K.O.; Konig, H.; Stackebrandt, E. *System. Appl. Microbiol.* **1983**, *4*, 535-551.
9. Brown, S.H.; Kelly, R.M. *Appl. Environ. Microbiol.* **1989**, *55*, 2086-2088.
10. Adams, M.W.W. *Microbiol. Rev.* **1990**, *75*, 219-238.
11. Mukund, S.; Adams, M.W.W. *J. Biol. Chem.* **1991**, *266*, 14208-16
12. Schafer, T.; Schonheit, P. *Arch. Microbiol.* **155**, 366-377.
13. Bragger, J.M.; Daniel, R.M.; Coolbear, T., Morgan, H.W. *Appl. Microbiol. Biotechnol.* **1989**, *31*, 556-561.
14. Blumentals, I.I.; Robinson, A.S.; Kelly, R.M. *Appl. Environ. Microbiol.* **1990**, *56*, 1255-1262.
15. Eggen, R.A.; Geerling, A.; Watts, J.; DeVos, W.M. *FEMS Microbiol. Lett.* **1990**, *71*, 17-20.
16. Connaris, H.; Cowan, D.A.; Sharp, R.J. *J. Gen. Microbiol.* **1991**, *137*, 1193-1199.
17. Brown, S.H.; Costantino, H.R.; Kelly, R.M. *Appl. Environ. Microbiol.* **1990**, *56*, 1985-1991.
18. Costantino, H.R.; Brown, S.H.; R.M. Kelly. *J. Bacteriol.* **1990**, *172*, 3654-3660.
19. Brown, S.H. Ph.D. Thesis, Johns Hopkins University, Baltimore, MD, **1992**.
20. Suzuki, Y.; Shinji, M.; Eto, N. *Biochim. Biophys. Acta.* **1984**, *787*, 281-289.
21. Suzuki, Y.; Yuki, T.; Kishigami, T.; Abe, S. *Biochim. Biophys. Acta.* **1976**, *445*, 386-397.
22. Ramsey, G; Freire, E. *Biochemistry* **1990**, *29*, 8677-8683
23. Yasunobu, K. T.; Tanaka, M. In *Iron-Sulfur Proteins;* Lovenberg, W., Ed., Academic Press, New York, 1973, Vol. 2; pp 27-130
24. Lovenberg, W. ; Sobel, B. E. *Proc. Natl. Acad. Sci. USA* **1965**, *54*, 193-199
25. Watenpaugh, K. D., Sieker, L. C., Herriott, J. R.; Jensen, L. H. *Acta Crystallog.***1973**, *B29*, 943-956
26. Seki, Y., Seki, S., Satoh, M., Ikeda, A.; Ishimoto, M. *J. Biochem.* **1989**, *106*, 336-341
27. Meyer, J., Gagnon, J., Sieker, L. C., van Dorsselaer, A.; Moulis, J.-M. *Biochem. J.* **1990**, *271*, 839-841
28. Woolley, K. J.; Meyer, T. E. *Eur. J. Biochem.* **1987**, *163*, 161-166
29. Bruschi, M. *Biochim. Biophys. Acta* **1976**, *434*, 4 - 17
30. Voordouw, G. *Gene* **1988**, *69*, 75-83

31. Shimizu, F.; Ogata, M.; Yagi, T.; Wakabayashi, S.; Matsubara, H. *Biochimie* **1989**, *71*, 1171-1177
32. Bruschi, M. *Biochem. Biophys. Res. Commun.* **1976**, *70*, 615-621
33. Hormel, S.; Walsh, K. A.; Prickril, B. C.; Titani, K.; LeGall, J.; Sieker, L. C. *FEBS Lett.* **1986**, *201*, 147-150
34. Bachmeyer, H.; Yasunobu, K. T.; Peel, J. L.; Mayhew, S. *J. Biol. Chem.* **1968**, *243*, 1022-1030
35. Bachmeyer, H.; Benson, A. M.; Yasunobu, K. T.; Garrard, W. T.; Whiteley, H. R. *Biochemistry* **1968**, *7*, 986-996
36. Saeki, K.; Yao, Y.; Wakabayashi, S.; Shen, G.-J.; Zeikus, J. G.; Matsubara, H. *J. Biochem.* **1989**, *106*, 336-341
37. Adman, E. T.; Sieker, L. C.; Jensen, L. H.; Bruschi, M.; LeGall, J. *J. Mol. Biol.* **1977**, *112*, 113-120
38. Watenpaugh, K. D.; Sieker, L. C. ; Jensen, L. H. *J. Mol. Biol.* **1979**, *131*, 509-522
39. Sieker, L. C.; Stenkamp, R. E.; Jensen, L. H.; Prickril, B.; LeGall, J. *FEBS Lett.* **1986**, *208*, 73-76
40. Frey, M.; Sieker, L. C.; Payan, F.; Haser, R.; Bruschi, M.; Pepe, G.; LeGall, J. *J. Mol. Biol.* **1987**, *197*, 525-541
41. Blake, P. R.; Park, J. B.; Bryant, F. O.; Aono, S.; Magnuson, J. K.; Eccleston, E.; Howard, J. B.; Summers, M. F.; Adams, M. W. W. *Biochemistry* **1991**, *30*, 10885-10891
42. Papavassiliou, P.; Hatchikian, E. C. *Biochim. Biophys. Acta* **1985**, *810*, 1-11
43. Giblin, M.; Kelly, C. T.; Fogerty, W. M. *Can. J. Microbiol.* **1987**, *33*, 614-618
44. Suzuki, Y.; Shinji, M.; Eto, N. *Biochim. Biophys. Acta* **1984**, *787*, 281-289
45. Suzuki, Y.; Yuki, T.; Kishigami, T.; Abe, S. *Biochim. Biophys. Acta* **1976**, *445*, 386-397
46. Yang, S.-J.; Zhang, S.-Z. *Enz. Engineer.* **1988**, *9*, 210-212

RECEIVED March 3, 1992

Chapter 4

Thermally Stable Urease from Thermophilic Bacteria

Kenneth Runnion[1], Joan Combie[1], and Michael Williamson[2]

[1]J. K. Research Corporation, 210 South Wallace, Bozeman, MT 59715
[2] U.S. Army Chemical Research, Development and Engineering Center, Detection Technology Division, Aberdeen Proving Ground, MD 21010–5423

Thermophilic microorganisms producing heat stable ureases have been isolated from the thermal waters of Yellowstone National Park in Wyoming. Conditions for optimized laboratory culturing were investigated as a means to achieve an economical methodology for the commercial production of an alternative stable enzyme used in clinical, industrial and environmental enzyme-based assays. Partially purified enzymes from ion exchange chromatography exhibited optimum activities within the pH range of 5.5 to 7.0 in citrate buffer. Although greater activities were expressed on increasing temperatures to 70°C, acceptable activities could be attained at 25°C making these enzymes attractive for an array of applications. Stability analysis at 50°C for 209 hours revealed 94% activity retention compared to 7% retention for the conventionally used Jack Bean urease.

Enzymes are frequently used as labels for signal generation in immunological and diagnostic assays, flavor mediators in food science and industry and detectors/ biosensors for toxicity and pollution monitorization (1). As a consequence of their proteinaceous origin, they are frequently the most unstable component of any system. Current enzymes and their associated labeled reagents are usually shipped in freezer packs and stored at 4°C or below. As enzyme-based instrumentation and diagnostics progress from tightly controlled clinical or research laboratories into more demanding conditions of the consumer and industrial markets, the acquisition of more stable enzymes is essential.

The development of colorimetric immunoassays in the early 1980's using the enzyme, urease as a label (2-4) has stimulated the more recent development of novel urease-based immunoassays using potentiometry with a silicon sensor (5) and fluorimetry with fiber optic technology (6). Although not optimal for all colorimetric immunoassays, the distinct color change produced by urease using bromocresol purple as an indicator (yellow to purple), gives this nonradiometric label an advantage over other existing enzymes. The substrate, urea, is safe, inexpensive and stable in solution (7). The absence of urease in most mammalian tissues reduces background noise yielding increased sensitivity (7). An appropriate enzyme label is selected in an effort to achieve the smallest size possible thus avoiding steric hindrance on binding of a conjugate to a complimentary antigen. Microbial ureases, as opposed to Jack bean urease which has a relative molecular mass (Mr) of 590K *(7, 8)*, are considerably smaller having been reported in the range of 125K to 380K *(7)*.

Commercially available urease is obtained predominantly from jack beans *(Canavalia ensiformis)*. Manufacturers suggest storage in a dry state at 4 to -20°C for stability. Such low temperatures are not practical for home health care kits, diagnostic kits in physician offices and health clinics and fieldable kits for monitoring ecotoxicity and pollution. Greater enzyme stability under a wider range of conditions is required.

Conventional means for increasing half lives of enzymes have included immobilization, entrapment, encapsulation and the addition of various stabilizing compounds. These techniques are often laborious and require the use of hazardous materials. Novel and more practical approaches to enzyme stabilization which impart actual changes to the amino acid sequence, have included direct isolation from thermophilic microorganisms, genetic engineering and manipulation *(9)* and active site surrogate construction via chemical synthesis *(10)*.

Characteristics of Thermophilically Derived Bacterial Enzymes

Microorganisms inhabiting extreme environments often produce markedly more stable enzymes, accounting for their ability to thrive under adverse conditions. More specifically, procaryotes (ie. bacteria as opposed to algae and fungi) are often preferred as sources of these enzymes due to their more frequent existence in more stressful thermal environments exceeding 70°C *(11)*. Thermally stable enzymes usually possess only small amino acid differences when compared to their mesophilic counterparts. However, these subtle changes often result in increased non covalent, intramolecular interactions and compaction sufficient to confer thermostability *(11)*. As a result of these additional interactions, thermostable enzymes often exhibit enhanced stability to other normally disruptive factors such as denaturation by detergents and organic solvents *(12)*. Although some mesophiles produce thermally stable enzymes, screening thermophiles is a more productive way to search for heat stable enzymes.

Geothermal activity is responsible for the creation of numerous moderate to high temperature environments, and thus serve as a rich source of thermophilic microorganisms. The Yellowstone region in northwestern Wyoming with more than 10,000 thermal features encompasses the largest and most varied array of geothermal phenomena on earth. Microorganisms producing heat stable urease were screened from these thermal waters..

Urease activity has been observed in numerous plants, bacteria, fungi, yeast and algae. Among soil bacteria, as many as 30% produce urease *(7)*. However, the occurence of urease in thermophiles has not been well documented. Bhatnagar did speculate on the existence of urease in Methanobacterium thermoautotrophicum albeit the enzyme was not identified or further characterized *(13)*. A Japanese patent on Bacillus sp. TB-90 is the only definitive reference in the literature on a urease produced by a thermophile *(14)*.

Many thermal waters are low in nitrogen, particularly organic forms. Unlike moderate temperature habitats, high temperature environments seldom contain any urea. In addition, other forms of fixed nitrogen are not abundant. Interestingly, only one small thermal basin in Yellowstone contains high concentrations of nitrogen (as much as 424mg/ L quantitated as ammonia) compared with most other sampling sites in the region with nitrogen concentrations below 5mg/ L *(15)*. For the thermophiles screened in this investigation, urease activity was not common.

Table I. Medium Composition For Organism Screening and
Isolation

Medium Components	Concentration (grams/ liter)
peptone	0.25
yeast extract	0.25
dextrose	2.0
magnesium sulfate	0.05
calcium chloride	0.02
sodium chloride	0.5
urea[a]	0.5
sodium phosphate[b]	1.0
Gelrite	8.0
manganese sulfate	0.001
zinc sulfate	0.000005
nickel chloride	0.00001

[a] filter sterilized - not used in controls

[b] monobasic or dibasic was used according to the desired pH

Collection and Isolation of Microorganisms

Water samples containing microorganisms were collected in Yellowstone National Park from thermal basins possessing a range of temperature and chemical characteristics. The selection of sampling sites primarily focused on previously recorded temperature, pH and nitrogen analysis data. More specifically, sampling sites exhibiting a temperature between 50 and 70°C, a pH near neutral and varied nitrogen concentrations were selected as the most promising sites for organisms producing urease enzymes of most commercial practicality. An attempt to increase the probability of recovering ureolytic organisms was accomplished using autosampler vials (Sun Broker, Inc., Wilmington, NC) containing various nitrogen sources. Rubber septa in the caps were replaced with porous membranes. These vials remained in thermal pools for 24 hours. Although no microbes were recovered from vials containing urea, a number of organisms were recovered from the same locations in vials containing either ammonium sulfate or potassium nitrate. Additional organism isolates from J. K. Research's Yellowstone thermophile collection were also screened for urease activity. Although previously collected for other enzymes, a few of the later organisms did possess apparent urease activity.

Within 8 hours of collection, water samples were returned to the laboratory and plated on medium specifically formulated based on the growth requirements of other thermophiles and ureolytic microorganisms (Table I) *(7, 11)*. The medium was adjusted to within one pH unit of the specific collection site. The cultures were incubated at the temperature most nearly approximating the original habitat (ie. 50 - 70 °C).

Enzyme Assay

Urease converts urea to ammonia and a carbamate intermediate. Carbamate spontaneously breaks down to a second ammonia and carbonic acid. At near neutral pH, carbonic acid dissociates and the ammonia becomes protonated, thus resulting in a net pH increase. This reaction provides the mechanism for screening, quantitation and validation of urease in documented methodologies including the indophenol, nesslerization, coupled enzyme, pH indicator and potentiometric assays *(7)*. In commercial and R&D assays incorporating a confirmed urease, a pH indicator such as bromocresol purple *(2)* or phenol red *(7)* with the substrate, urea, is most frequently used due to simplicity. The substrate and dye are added to appropriate buffers containing the enzyme. A pH increase yielding variation in color is indicative of urease activity and is monitored by eye or quantitated spectrophotometrically.

Selection of Assay Methodology. In contrast to other more specific urease detection methods, the pH indicator assay is simple and provides the necessary rapidity for screening numerous isolates from an enviroment generally non-conducive to urease producing organisms. Although non conventional for confirmational urease analysis, the pH indicator assay can be used with reasonable confidence given the incorporation of appropriate controls and validation assays. A modified pH indicator assay, substituting a standard micro pH electrode for the indicator, was used throughout this investigation. This method allowed several media in the screening process and buffers in later quantitation and characterization assays to be used with different initial pH's. A change in pH recorded as Δ pH units was calculated as:

$$\Delta \text{ pH Units } = \left[-\log [H^+]_{T_{endpoint}} - (-\log [H^+]_{T_0}) \right] \quad \text{Where T = assay time}$$

In the crude screening process, organisms were streaked on plates in the presence and absence of urea. A pH increase of greater than 1 unit in the media containing urea as compared with comparable media not containing urea was considered positive for presumptive urease. Of two hundred twenty five cultures screened, thirty yielded positive for apparent urease activity.

Validation of presumed urease production was initially accomplished by monitoring liberated ammonia using Merck's EM Quant test strips. Only eight of the thirty isolates producing urease by this assay were deemed to possess sufficient activity to warrant further investigation and quantitation. These organisms originated from habitats ranging in temperature from 47 to 75°C and pH from 5.4 to 8.1. Only one isolate (#429) with enhanced thermophilic characteristic (demonstrated by an optimum growth temperature at 70°C) would produce detectable urease activity at 70°C. Further confirmation of urease production while differentiating from the frequently observed mesophilic ATP-urea amidolyase (UALase) was accomplished by monitoring urease inhibition in the presence of hydroxyurea *(7, 16,17)*. UALase has been observed in yeast and green algae while urease is usually observed in urea degrading bacteria. As similarly observed by Mackeraas and Mobley, an average of 80% and 45% inhibition in the presence of 0.00076 and 0.00009 gm/ L hydroxyurea, respectively, was confirmational that urease but not UALase existed in all the isolates assayed *(7,17)*.

All of the data described in the text was gathered from relative quantitation and characterization assays. These assays were performed by withdrawing an aliquot of sample containing the enzyme from the culture medium at selected intervals or from semi-purified preparations. Occasionally, the aliquot was diluted and one mL mixed with 2 mL of selected buffer. Following the determination of activity-pH profiles for each isolate, citrate buffer at 50 mM (pH 7.0) was observed to be optimum and thus used for further characterization. Immediately following the addition of one mL of a stock concentration of urea at 25 gm/ L, the pH was monitored at a predetermined time and temperature (usually 1 hour or overnight at 20°C). Appropriate controls containing organisms without urea exhibited no pH change.

Presence of Stabilizers. The inactivation of urease via the oxidation of labile sulfhydryls in the presence of oxygen and heavy metals has frequently necessitated the incorporation of oxygen scavengers, metal chelators and other stabilizing agents (ie. reductants) to the assay medium *(18,19)*. Consistent with most other investigations, a concentration of 1 mM of EDTA was used throughout this investigation to chelate and minimize the availability of heavy metals.

The incorporation of various thiol reducing compounds such as dithiothreitol, glutathione and 2-mercaptoethanol have also been routinely used in assay buffers to further protect the delicate sulfhydryls. This practice has recently been significantly reduced since such compounds have been observed to exhibit significant inhibition at high concentrations. However at lower concentrations, the effect has been minimal. Our limited experience with 2-mercaptoethanol on thermally stable ureases has yielded similar results and therefore was omitted from the assays.

Optimization of Culture Conditions

Nitrogen source. Regulation of urease biosynthesis may be repressible, inducible or constitutive *(7, 20)*. In many bacterial species, urease biosynthesis is controlled in conjunction with the nitrogen regulatory system. Synthesis may be derepressed under nitrogen limiting conditions or induced by the presence of urea. For organisms in which urease production is constitutive, enzyme activities are unaffected by the addition or limitation of nitrogenous compounds *(7)*.

Regardless of substantial growth in our investigation, significant urease repression in all isolates in the presence of urea was particularly noted and may partially explain the previously observed absence of organisms collected in membrane enrichment vials containing urea. Bast observed similar correlation of urease production with ammonia concentration where urease biosynthesis oscillated with varying concentrations of exogenous ammonia *(21)*. Mobley has also observed repression of urease synthesis by urea *(7)*. Although urease induction by urea can occur, the mechanism has been suggested to be secondary and less effective for urease synthesis. Table II illustrates the repression of urease by urea in 6 organisms when urease activity is monitored in the presence of various nitrogen sources. Although biomass production was not inhibited, the presence of urea at 1 gm/ L as the predominant sole nitrogen source resulted in significantly less urease activity in all 6 organism isolates. Other urea derivatives, ie. uric acid and hydroxyurea, were also similarly observed to promote urease repression (Data not shown). The later, however, may be more related to urease inhibition.

Table II. Effect of Nitrogen Source on Urease Production

Organism Isolate #	Peptone (0.1)	Peptone (0.25)	Peptone (1)	Peptone (3)	NH_4Cl (5)	Urea (1)
197	1.57[a]	0.67	0.57	0.17	0.82	0.12
211	1.97	0.94	0.56	0.15	0.24	0.10
288	1.58	1.61	1.44	0.88	1.91	0.10
301	1.03	0.88	0.93	0.14	1.26	0.07
369	2.02	2.07	1.94	0.59	2.13	0.13
401	1.23	1.57	1.78	0.71	1.53	0.14

Nitrogen Source (gram/ Liter)

[a] Relative urease activities expressed as a change in pH units.
Cultures were raised on medium (Table I) modified by substituting the indicated nitrogen sources for peptone and urea. Following two days of growth at 50°C, 1 mL aliquots of cell suspensions of equivalent concentrations were quantitatively assayed in 2 mL of HEPES buffer (0.2M, pH 7.5) containing 100 mM urea at ambient temperature overnight. Appropriate assay controls in the absence of urea exhibited no change in pH.

Organism isolate #429, with enhanced thermophilic characteristic also exhibited similar urease repression when urea was used as a nitrogen source (Table III). The apparent stimulation of urease by casein might be associated with nitrogen limitation as could be inferred from the recent demonstration of the casein fraction of milk (human) contributing less than 25% of the total available nitrogen (22).

Table III. Effect of Nitrogen Source on Urease Activity in Isolate #429

Nitrogen Source	Concentration (grams/ Liter)	Activity [a]
casein	1.0	1.09
tryptone	0.5	0.79
urea	0.5	0.17
ammonium sulfate	0.5	0.15

[a] Relative urease activity expressed as a change in pH units.
Organisms were cultured and enzymes assayed analogously to that described in Table II.

Dextrose. Dextrose is frequently used as a sole carbon source in minimal media for laboratory culturing of numerous microorganisms. Although urease repression by dextrose has been previously reported (23), we observed a significant increase in urease activity as dextrose concentrations were elevated to 25 gm/ L (Table IV). This increase is further correlated with biomass accumulation which is consistent with other investigators who observed increases in urease activity during the early stationary phase or late log phase (20). Later results on optimization revealed additional increases in biomass production when casein was substituted for peptone. This increase might perhaps be synergistic with nitrogen limitation as previously discussed toward urease stimulation or biosynthesis. Casein digests and hydrolysates are known to induce the LAC operon resulting in accelerated carbohydrate fermentation and increased biomass. The growth of other organisms such as salmonellae have similarly been observed to be stimulated by casein *(22)*. As a consequence of increased urease activity in our isolates, casein was substituted for peptone in the final media formulation.

Table IV. Effect of Dextrose Concentration on Urease Production

Organism	Dextrose Concentration	
Isolate #	2 gm/ L	25 gm/ L
197	1.16[a]	2.73
211	1.84	2.67
288	1.19	2.45
301	0.85	2.13
339	1.26	2.26
369	0.57	1.52

[a] Relative urease activity expressed as a change in pH units.
Organisms were cultured and enzymes assayed analogously to that described in Table II.

Other Nutrients. A final medium formulation was constructed for future large scale fermentation and optimized urease production (Table V). Although urease is a nickel containing enzyme and this element was added to the screening media in Table I, the addition of nickel was later observed unnecessary presumably because sufficient trace quantities were being obtained from other components. The addition of other micronutrients including salts of iron, manganese, zinc, copper, cobalt and molybdate were added similarly as in other micronutrient mixtures used for the growth of other thermophiles *(24)*. Although zinc sulfate at 0.01 gm/ L was inhibitory, the element proved beneficial to some isolates at 0.001 gm/ L while being not required in others. Magnesium sulfate had no detrimental effect to a maximum of 0.25 gm/ L, however some decrease in urease activity was generally observed at 0.5 gm/ L. TRIS buffer generally reduced urease activities while sodium or potassium phosphate proved optimum. The rather obscure data observed on micronutrient manipulation might be related to the inclusion of casein. While optimal enzyme activity was observed with 1.0 gm/ L casein, reducing the concentration to 0.1 gm/ L or raising it to 5 gm/ L generally yielded 80-90% lower activities. Casein with an average molecular weight of 25,000 is composed of several protein subclasses (α, β, κ and γ) known to form complex micelles that bind a number of metals and cations including zinc, magnesium, manganese, iron, calcium, copper and phosphate *(22)*. This binding has the potential for limiting bioavailability especially as casein concentration increases. However, a potential advantage of casein is suggested by a recent report that it is able to bind the heat stable enterotoxin produced by E. coli and this binding is more pronounced in acidic environments *(22)*. This phenomenon and the reduction of pH as fermentation proceeds has the potential for prolonging the growth of bacteria.

Table V. Final Medium Formulation For Optimized Urease Production

Medium Components	Concentration (grams/ liter)
yeast extract	0.5
dextrose	25.0
casein	1.0
magnesium sulfate heptahydrate	0.1
sodium phosphate dibasic	3.0
micronutrients [a]	*

[a] micronutrients at 1.0 mL/ L of a final preparation containing 0.29 gm/ L ferric chloride hexahydrate, 2.28 gm/ L manganese sulfate monohydrate, 0.40 gm/ L zinc sulfate heptahydrate, 1.50 gm/ L boric acid, 0.05 gm/ L copper sulfate pentahydrate, 0.20 gm/ L molybdic acid sodium salt, and 0.05 gm/ L cobalt chloride hexahydrate (24).

Aeration. The effect of air on urease activity during culturing was examined in 3 day organism cultures that had been magnetically stirred at 400 rpm (minimal aeration), shaken in a waterbath at 80 rpm (intermediate aeration) and sparged with air at a rate of 1 mL air/ mL medium/min (maximal aeration). Aeration via magnetic stirring is perceived to produce less oxygenation than shaking the entire flasks. Table VI illustrates maximum urease activity with maximum aeration of the 6 organism isolates investigated. Although, the significant increase in urease activity observed at higher aeration rates might simply represent the organism's response to an increased nitrogen availability, an increase in biomass with subsequent increased urease yields as observed in the presence of added dextrose, is also possible.

Table VI. Effect of Air on Urease Production

Organism Isolate #	Minimal	Relative Aeration[a] Intermediate	Maximal
197	0.11[b]	0.28	2.59
211	0.11	0.66	2.18
288	0.05	0.55	0.86
301	0.12	0.37	2.11
368	0.00	0.00	2.27
401	0.12	0.20	1.82

[a] See text for definition
[b] Relative urease activity expressed as a change in pH units.
Organisms were cultured and enzymes assayed analogously to that described in Table II.

Urease Purification

Cell Disruption. Prior experience with mesophiles has generally shown that urease is a cytoplasmic enzyme. However some evidence that the enzyme may be membrane bound or possibly extracellular exists (7). In this investigation, only 20% of the urease activity obtainable after sonication could be recovered in the culture media following simple centrifugation of the intact organisms. However, these results varied with the organism and culture parameters. Some means of cell disruption was found necessary for efficient harvesting of the enzyme being retained by the cell. Of the techniques attempted, freeze thawing and nitrogen cavitation were less efficient than sonication with multiple freeze thawing yielding practically 9 fold less than a 5 minute sonication. Optimal conditions for sonication included 45 minutes for 1 liter of cell suspension using a 600 watt sonicator with a one inch horn. The duration of sonication for complete cell disruption varied among organism, temperature and sample size owing to the robust characteristic of the thermophile. Cellular debris was removed and the protein fraction concentrated using an ultrafiltration system by Millipore.

Chromatography. Affinity chromatography using either urea or hydroxyurea is commonly attempted by many investigators to enhance the specificity for urease. However, as similarly observed by some investigators *(25-27)*, our efforts to parallel this procedure demonstrated the approach as non viable. Other attempts to use various dye ligands as affinity probes were also unsuccessful in purification. The enzymes were observed not to bind to Amicon dyes including red A, green, orange, blue A, and blue B in four buffers ranging from pH 5.5 to 7.8. Athough binding to the columns could not be achieved, eluted urease from the green dye exhibited enhanced stability at ambient temperature suggesting the binding and removal of some inhibitor and/or protease.

Most protein purification procedures use anion exchangers such as DEAE *(19, 28-30)*. In this investigation, a strong anion exchanger, Macro-Prep 50Q from Bio-Rad, was observed to yield superior results. The cell-free extract was applied to the column in 50 mM HEPES buffer containing 1 mM EDTA at pH 7. The column was washed with the same buffer followed by an analogous buffer containing 0.1 M sodium chloride to remove weakly bound contaminants. Thermally stable urease was eluted in HEPES buffer (50mM) containing 0.5 M sodium chloride. The eluted urease was desalted by dilution using a tangential flow concentrator system. The entire process was performed at ambient temperature.

Characterization of Thermally Stable Urease

Optimum pH. The optimum pH for ureases is well established to be strongly buffer dependent which may be related to inhibitions caused by buffer composition. Ureases have been observed to be inhibited by sodium, potassium and ammonium cations and phosphate anions especially at pH's less than 7. Nakano observed optimal pH's of 6.2, 7.4 and 7.3 for maleate, Tris-HCl and phosphate buffers, respectively *(19)*. Additionally, Jespersen and Taylor observed optimal pH's ranging from 6.5 to 8.0 for phosphate, maleate, citrate and tris buffers. The relative amount of activity also varied with the buffer even at the optimal pH *(31, 32)*. Only 42% of the urease activity in maleate buffer was observed at the optimal pH in phosphate buffer *(19)*. Similar results can be observed in Figure 1 which depicts the pH - activity profiles for 3 organism isolates. These profiles were performed with an excess of urea substrate so as to prevent interferences from alterations in Km. Rather high buffer concentrations were also used so as to impede major pH changes which would otherwise result in artefactual phase shifts of the profiles. For all three isolates examined, citrate buffer proved to be superior yielding as much as 3 fold greater activities. In contrast, no activities were exhibited in phosphate buffer for isolates 429 and 197 at pH's less than 7 as previously documented with other ureases *(32)*.

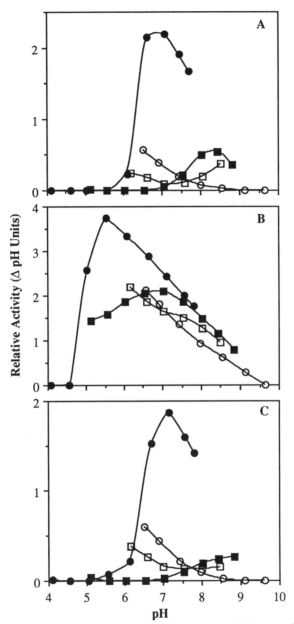

Figure 1. Activity-pH profiles for Isolates 429 (A), 408 (B) and 197 (C) in Citrate ●, Phosphate ■, Tris ○, and HEPES □ buffers. Aliquots (0.1 mL) of partially purified enzyme from DEAE chromatography were assayed in 3 mL of 200 mM buffers containing 136 mM urea for 15 hours at 20°C. Relative urease activities were calculated and recorded as a change in pH units.

Optimum Temperature. Table VII illustrates the effect of temperature on the kinetics of urease activity from isolate 413. Consistent with many other enzymes from thermophiles, urease activity generally increases with increasing temperatures to 60°C. The relative proportional increases in activities with increasing temperatures particularly during first order kinetics is presumably due to the expulsion of inhibitory ammonium ions as observed by Magana-Plaza *(18)*. A substrate limiting condition which results in a pH stabilization (mixed or zero order kinetics) occurs earlier as the temperature of the enzyme reaction is increased (Table VII).

Table VII. Temperature - Kinetic Profile of Urease from Isolate #413

Time	Temperature (°C)			
(Hours)	22	37	50	60
0.167	0.05a	0.17	0.40	0.64
0.500	0.16	0.67	1.08	1.21
1.250	0.53	1.44	1.52	1.53
2.000	1.06	1.69	1.64	1.64
2.500	1.45	1.77	1.73	1.71

a Relative urease activities, corrected for urea autohydrolysis at elevated temperatures, are expressed as a change in pH units.
Aliquots (1 mL) of partially purified urease from DEAE chromatography were assayed in 2mL of HEPES buffer (pH 7.5) containing 100 mM urea and 1mM EDTA at the indicated temperatures. At the indicated time intervals, the pH was monitored and Δ units calculated. Appropriate controls without urea exhibited no change in pH.

The relative activity-temperature profiles for 3 isolates with respect to buffer is illustrated in Figure 2. Although this is not a rate-thermal profile, some indication of thermal stability is portrayed. In contrast to HEPES buffer, ureases from isolates 408 and 197 in the presence of citrate buffer exhibit enhanced stability at 70°C by the end point method. Analogous to Table VII, the pH stabilization, particularly in the presence of citrate buffer (open symbols), is presumably due to substrate limitation as incubation time progresses. Although not confirmatory, the increasing activities of isolate 429 (Figure 2A) with increasing temperatures is highly suggestive of greater thermophilicity.

The ability to express activity above 50°C is characteristic of thermophilic as opposed to mesophilic enzymes since most mesophilic enzymes are known to rapidly denature above 50°C. Except in the presence of HEPES buffer, all isolates exhibit rather high activity at 70°C indicating thermophilicity. Suppressed activity of isolates 197 and 408 in HEPES at 70°C is unlikely indicative of thermal instabilility but rather to buffer inhibition since the enzyme appears stable in citrate at similarly elevated temperatures.

Specific Activity. Specific activities of enzymes, usually reported as units of activity per mg of protein, are used as comparative indicators of enzyme preparation quality and quantity. Increases in purity are reflected by subsequent increases in specific activity. The range of specific activities are extremely varied among ureases from various sources and are highly dependant on the choice of buffer, pH, solvent components and assay temperature. Although <u>Ureaplasma urealyticum</u> has been observed to possess specific activities ranging from 33,000 to 180,000 μmoles of urea/ min/ mg protein, highly purified microbial ureases are more commonly reported in

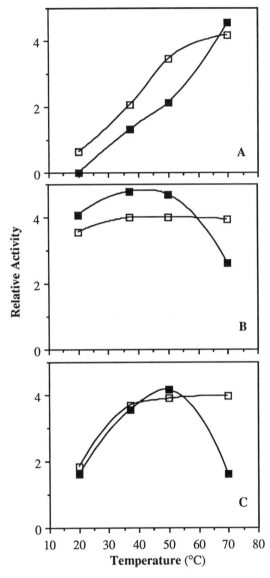

Figure 2. Relative activity-temperature profiles for ureases from organism isolates 429 (A), 408 (B) and 197 (C). Aliquots (0.1 mL) of partially purified enzyme from DEAE chromatography were assayed in 3 mL of 50mM citrate (open symbols) and HEPES (solid symbols) buffers containing 136 mM urea. Activities were calculated and recorded as a change in pH units by the endpoint method at 7.5 hours and normalized to organism 408 in citrate buffer.

the range of 9 to 5500 μmoles of urea/ min/ mg protein *(33)*. However a few have
been documented to have significantly lower specific activities in the range of 0.6-15
μmolar units/ mg protein *(18, 19, 25, 28, 29, 34)*. In contrast, the plant urease from
Jack Beans has been reported to exhibit a specific acitivity of 3500 μmolar units/ mg
protein at 37°C and pH 7. Sigma Chemical Co. distributes a Jack Bean urease with
approximately 780 μmolar units/ mg solid at 25°C and pH 7. Among bacterial sources, only Ureaplasma urealyticum exhibited comparable specific activities *(35)* to Jack
Bean urease (27).

In this investigation, a preliminary attempt to obtain specific activities was initiated using freshly reconstituted commercial Jack Bean urease from Sigma Chemical
Co. (type VII containing 6000 μmolar units per 8.5 mg solid where 1 μmolar unit is
equivalent to 1 μmole of ammonia liberated per minute at 25°C). A standard curve
was constructed using a serially diluted Jack Bean preparation. Aliquots from 10 to
100 μl were assayed in 2 mL of 5 mM phosphate buffer containing 1 mM EDTA,
100 mM urea, 0.05% BSA and 0.001% phenolphthalein adjusted to pH 8.0. First order catalytic rates were monitored spectrophotometrically at an absorbance of 555nm
at 25°C. According to the pH profiles for three isolates (Figure 1), the common and
near optimal 50 mM citrate buffer containing 1 mM EDTA, 100 mM urea and
0.001% phenol red adjusted to pH 7 was used to obtain similar catalytic rates at an
absorbance of 560 nm and a temperature range of 25-70°C for the thermostable
ureases. Controls containing urea but no urease were performed at individual temperatures and subtracted from the experimentals. The specific activities in Table VIII
are expressed as Jack Bean (JB) μmolar units. Although the specific activities of the
semi-purified, cell-free extracts are comparable to other reported bacterial ureases,
considerably enhanced activities are likely to be achieved with further investigation
and refined purification strategies. The data are further consistent with Table VII
which suggest increased thermophilicity of urease #429 as compared to #408 and 197
(ie. Increased thermophilicity is indicated by increased activity ratios at 70 and
25°C).

For optimum performance and catalytic efficiency, information on kinetic parameters such as the Michaelis Menten constant (Km) and maximal velocity (Vmax) are
required for enzymes used in analytical procedures. Kinetic estimates were obtained

Table VIII. Specific Activities for Thermostable Ureases

Enzyme Origin	Temperature (°C)			
(Isolate #)	25	37	50	70
429	.304[a]	.478	1.22	16.73
408	8.7	16.53	25.85	39.94
197	3.44	4.93	7.9	10.63

[a] relative activities of partially purified thermostable ureases from
DEAE chromatography are expressed as JB μmolar units per mg protein and established using a reaction rate method and extrapolated from
a standard curve (See text). Assays were conducted at least in qudrupli-
cate for 1 hour in 50 mM citrate (pH 7), 1 mM EDTA, 100 mM urea
and 0.001% phenol red as an indicator. Activities were calculated from
linear first order catalytic rates.

using a similar procedure where reaction rate was monitored as a function of substrate concentration between 0.25 mM and 333 mM in 50 mM citrate, 1 mM EDTA and 0.001% phenol red adjusted to pH 7. The suitability of the procedure was previously verified using Jack Bean urease in which Km's of 3.5 - 4.0 were achieved and closely approximates the 2.9 value reported by Blakeley *(7, 33, 36)*. The Km constants for thermostable ureases were experimentally determined at 5-15 mM for organism isolates #408 and 197 and considerably lower at 0.2-0.6 mM for organism isolate #429 dependant on the temperature of assay. An observed increase in Km at elevated temperature is consistent with that observed by Magana-Plaza who correlated a similar increase with an inhibition by ammonia *(18)*. The inhibition was further observed to coincide with a change from Michaelis Menten kinetics to a sigmoidal type profile. This effect was particularly evident with isolate 429 and was greater as the substrate concentration was progressively increased above 10 mM. At substrate concentrations less than 10 mM, the kinetic profile exhibited a linear Michaelis Menten mechanism. For bacterial ureases, Km values are generally reported in the range of 10-50 mM and less frequently as low as 0.1 and as high as 100 mM *(7, 33)*. Considering the extreme thermophilic nature of isolate #429 (activity optimum above 60°C) and the limited data on other thermostable ureases, uncommonly low reported Km's may not be so unusual.

Electrophoresis Several enzymes from organisms of moderate to higher thermophilic characteristic were electrophoresed on non-denaturing 7.5% polyacrylamide gels. Observed molecular weights ranged from over 200K for moderate thermophilically derived ureases to 120-130K for enzymes from organisms possessing enhanced thermophilic characteristics. Most bacterial ureases have been observed to possess molecular weights in the range of 200 to 380K, however Ureaplasma urealyticum has been reported as low as 150K and as high as 380K *(7)*. An unidentified bacterium has been reported to synthesize a urease of 125K *(7)*.

Stability. The fact that many thermophilic bacteria from hot springs remain active to temperatures exceeding 100°C *(11)*, provides the basis for belief that their enzyme products must also be stable at the temperature approximating the habitat. For most ureases isolated from mesophilic organisms, stability at 0°C is monitored by month. Commercially available Jack Bean urease is reportedly stable for 1 year at minus 20°C *(7)*. Less data is available at higher temperatures. Urease from Proteus mirabilis and Selenomonas ruminantium are stable for 10 minutes at 60°C *(7)*. The Bacillus sp. TB-90 isolated by the Japanese produce a urease which retains full activity following incubation for 17 hours at 37°C or 15 hours at 50°C *(14)*. Attempts to increase the stability of urease have included buffer formulations, bacteriostats, chelating agents *(37)*, glycerol *(19)* and immobilization *(1)*.

Binding to aminopropyl glass beads was unsuccessful. Although partially purified urease from isolate 408 bound at a low rate to arylamine glass, that which was bound retained nearly full activity for 48 hours when stored dry at 50°C.

Table IX illustrates the relative stabilities of enzymes from isolates 408 and 429 as compared with Jack Bean after almost 9 days of storage. The stability analyses were conducted in glycerol to allow a better comparison with the Jack Bean enzyme which is predominantly commercially distributed in the presence of glycerol among other additives. As a consequence of the uncertainty of the commercial urease composition, accurate stability comparison is difficult and the resultant stability ratio between the thermostable and Jack Bean is likely to be much greater. In a similar stability on a crude preparation from isolate #429, retention of 95% activity after 40 hours at 60°C was observed.

Table IX. Stabilty Comparison of Thermostable and Jack Bean Urease

Enzyme	Storage Temperature	
	37°C	50°C
#408	103a	90
#429	98	94
Jack bean	46	7

aRelative percentage activity remaining following the storage of partially purified #408 and 429 enzymes from DEAE chromatography at the indicated temperatures for 209 Hours. All three enzymes were stored in 50% glycerol containing 1mM EDTA. Jack Bean urease was obtained from Sigma Chemical Co. Aliquots (0.25 mL) were assayed in citrate and HEPES buffers at 50 mM (pH 7) for thermostable enzymes and Jack Bean urease, respectively. Assays were conducted at 37°C for #408 and Jack Bean urease and 50°C for #429. Activities were calculated by endpoint at 2 hours.

Applications and Summary

The most common application of urease is perhaps for laboratory determination of urea in foodstuffs and biological fluids (26). Analytical instruments sensing a change in surface potential (38) and visually monitored ELISA's frequently use the enzyme as a label to generate a desired signal.

The commercial expansion of urease use has more recently been broadened outside the laboratory. A patent, issued in California in 1989 for a preservative of harvested crops, listed urease as the predominant active ingredient. The principle was based on sufficient ammonium ions being generated to inhibit bacterial growth (39). The Japanese were granted a patent for use of urease to remove carbamide from alcoholic liquors. The formation of ethyl carbamate, a carcinogen, is prevented by using urease to eliminate urea (40). Additionally, urease sensitivity to a variety of metals has resulted in the successful implementation of an immobilized urease as a probe for the detection of metal contaminants in the environment (16).

Although numerous industrial, academic and domestic applications exist for ureases, novel applications are continually being developed. A more stable enzyme exhibiting reduced inhibitory restrictions would undoubtedly find expanding markets in addition to improving existing technology. For example, ureases are often components of complex inoculum mixtures used in waste reduction and composting procedures. The ureases function in intermediary and precursor reactions to a much larger series of enzyme cascade reactions. As a result of urease instability, composting is limited to mild, temperate climates and continual innoculum refurbishments and urea substrate application (41, Rutherford, J.P., Harford County, MD Landfill, Personal Communication, 1990). The thermostable ureases described here have great potential for increasing the efficiency and productivity of processes commonly plagued by unstable enzymes. Several organism isolates possessing ureases of varying degress of stability and catalytic characteristics have been described. The unique region of geothermal activity and chemical composition from which the organisms were obtained is presumably a significant factor to the enzyme characteristics observed. Although much additional investigation is required, the specific activities, stabilities and kinetic parameters already suggest significant superiority to the practically universally used Jack Bean urease.

Legend of Symbols

Δ	Change In
T_o	Time of Initial Measurement
$T_{endpoint}$	Time of Final Measurment

Acknowledgements

This work was supported through a grant (Contract #DAAA-15-89-C-O501) from the US Government Small Business Innovation Research (SBIR) Program administered through the US Army Chemical Research, Development and Engineering Center (CRDEC) at Aberdeen Proving Ground, MD USA.

Special thanks to the United States National Park Service for permitting us to collect samples from Yellowstone Park, Montana. The co-operation of all Park Service personnel contributed to the success of our research. We would like to specifically mention the assistance provided by Rick Hutchison, District Rangers John Lounsbury, Joe Evans, Jerry Mernin and Steve Frye and Research Administrator, John Varley.

Literature Cited

1. Guilbault, G.C. In *Analytical Uses of Immobilized Enzymes*, Marcel Dekker Inc., **1984**
2. Chandler, H.M.; Cox, J.C.; Healey, K.; MacGregor, A.; Premier, R.R.; Hurrell, J.G.R. *J. Immunological Meth.* **1982**, *53*, 187-194.
3. Lo, C.Y.; Notenbohm, R.H.; Kajioka, R. *J. Immunol. Meth.* **1988**, *114*, 127-137.
4. Meyerhoff, M.E.; Rechnitz, G.A. In *Methods in Enzymology*; Editor, H. Van-Venakis; J. Langone; *Vol 70*; Academic Press: **1980**; pp 439-454.
5. Olson, J.D.; Panfili, P.R.; Armenta, R.; Femmel, M.B.; Merrick, H.; Gumperz, J.; Goltz, M. and Zuk, R.F. *J. Immunol. Meth.* **1990**, *134*, 71-79
6. Rhines, T.D. and Arnold, M.A. *Analytica Chimica Acta* **1989**, *227*, 387-396.
7. Mobley, H.L.T.; Hausinger, R.P. *Microbiol. Rev.* **1989**, *53*, 85-108.
8. Takishima, K.; Suga, T.; Mamiya, G. Eur. *J. Biochem.* **1988**, *175*, 151-165.
9. Perry, L.J. and Wetzel, R. *Science (Wash DC).* **1984**, *226*(4674), 555-557
10. Twork, J.N. and Yacynych, A.M., *Bioprocess. Technol.* **1990**, *6*.
11. Brock, T.D. In *Thermophiles*; John-Wiley & Sons: New York, NY, **1986**.
12. Sharp, R.J.; Munster, M.J. In *Microbes in Extreme Environments*; Editor R.A. Herbert; G.A. Codd; Academic Press; London, **1986**, p 233.
13. Bhatnagar, L.; Jain, M.K.; Aubert, J.-P.; Zeikus, J.G. *Appl. Environ. Microbiol.* **1984**, *48*, 785-790.
14. Takashio, M.; Yoneda, Y.; Mitani, Y. U.S. Patent 4,753,882, **1988**.
15. Thompson, J.M.; Presser, T.S.; Barnes, R.B.; and Bird, D.B. US Geological Survey open file report #75-25, **1975**.
16. Rai, A.K.; Singh, S. *Curr. Microbiol.* **1987**, *16*, 113-117.
17. Mackerras, A.H.; Smith, G.D. *J. Gen. Microbiol.* **1986**, *132*, 2749-2752.
18. Magana-Plaza, I.; Montes, C.; Ruiz-Herrera, *J. Biochim. Biophys. Acta* **1971**, *242*, 230-237.
19. Nakano, H.; Takenishi, S.; Watanabe, Y. *Agric Biol. Chem.*, **1984**, *48*, 1495-1502.
20. Bast, E. *Arch. Microbiol.* **1986**, *146*, 199-203
21. Bast, E. *Arch. Microbiol.* **1988**, *150*, 6-10.
22. Miller, M.J.S.; Witherly, S.A. and Clark, D.A. In: *Proceedings of the Society for Experimental Biology and Medicine*, 195(2), **1990**, 143-159.

23. Yuodval'kita, D.Y.; Glemzha, A.A.; Gal'vidis, I.Y. *Microbiol. (Moscow)* **1982**, *51*, 919-925.
24. Castenholz, R. and Pierson, B.; *The Procaryotes* , Eds. Starr, M.; Stolt, H.; Truper, H.; Balow, A. and Schleger, H., **1981**, 1, 290-298.
25. Creaser, E.H.; Porter, R.L. *Int. J. Biochem.* **1985**, *17*, 1339-1341.
26. Mendes, M.J.; Karmali, A.; Brown, P. *Biochimie*, **1988**, *70*, 1369-1371.
27. Wong, B.L.; Shobe, C.R. Can. *J. Microbiol.* **1974**, *20*, 623-630.
28. Breitenbach, J.K.; Hausinger, R.P. *Biochem. J.* **1988**, *250*, 917-920.
29. Dunn, B.E.; Campbell, G.P.; Perez-Perez, G.I.; Blaser, M.J. *J. Biol. Chem.* **1990**, *265*, 9464-9469.
30. Hu, L-T.; Nicholson, E.B.; Jones, B.D.; Lynch, M.J.; Mobley, H.L.T. *J. Bacteriol.* **1990**, *172*, 3073-3080.
31. Jespersen, N.D. *J. Am. Chem. Soc.* **1975**, *97*, 1662-1667.
32. Taylor, M.B.; Goodwin, C.S.; Karim, Q.N. *FEMS Microbiol. Lett.* **1988**, *55*, 259-262.
33. Mobley, H.L.T.; Cortesia, M.J.; Rosenthal, L.E.; Jones, B.D. *J. Clin. Microbiol.* **1988**, *26*, 831-836.
34. Todd, M.J.; Hausinger, R.P. *J. Biol. Chem.* **1989**, *264*, 15835-15842.
35. Saada, A-B.; Kahane, I. *Zbl. Bakt. Hyg.* **1988**, *A269*, 160-167.
36. Blakeley, R.L.; Webb, E.C. and Zerner, B. *Biochemistry* **1969**, 8, 1984-1990.
37. Modrovich, I.E. U.S. Patent 4,378,430, **1980**.
38. Briggs, J.; Kung, V.T.; Gomez, B.; Kasper, K.C.; Nagainis, P.A,; Masino, R.S.; Rice, L.S.; Zuk, R.F.; Ghazarossian, V.E. *BioTechniques* **1990**, *9*, 598-606.
39. Young, D.C. U.S. Patent 4,822,624, **1989**.
40. Kakimoto patent 4,844,911, **1989**.
41. Rutherford, J.P., *Agricultural Innoculum Composition*, US Patent #4,337,077, 29 June **1982**

RECEIVED January 15, 1992

Chapter 5

Respiratory Electron-Transport Components in Hyperthermophilic Bacteria

R. J. Maier, L. Black, T. Pihl[1], and B. Schulman[2]

Department of Biology, The Johns Hopkins University, Baltimore, MD 21218

The hyperthermophilic anaerobic archaebacterium Pyrodictium brockii grows optimally at 105°C by oxidizing H_2 with elemental sulfur to form H_2S. The membrane bound H_2-oxidizing hydrogenase has properties that are strikingly similar to the H_2-uptake hydrogenases from (eubacterial) aerobic H_2 oxidizers. These properties include metal content, molecular subunit properties, O_2 inhibition properties, and artificial electron acceptor specificity. The purified P. brockii enzyme is still rather sensitive to thermal inactivation (loss of 70% of activity in 15 min at 90°C), but the membrane bound form can with-stand 90°C for 15 min with only a 20% activity loss. The sequential electron carriers downstream of hydrogenase also have similarities to aerobic H_2 oxidizing electron transport chains. A quinone and a c-type cytochrome were identified in P. brockii membranes, and an H_2 to sulfur electron transport pathway could be restored by adding the purified quinone to UV-light treated membranes. Heme staining of SDS gels revealed a single heme-staining component (c-type cytochrome) of molecular mass 13-14 kDa. The cytochrome probably plays the unusual function of a quinol oxidase. The limited available energy in the coupling of H_2 oxidation to S° reduction may constrain the number of components permitted to efficiently capture energy in this presumably primitive electron transport chain.

Our work has involved the characterization of the H_2-uptake hydrogenase and the associated respiratory chain in the chemolithotrophic thermophile Pyrodictium brockii. The reader is referred to the work of

[1]Current address: Department of Microbiology, The Ohio State University, Columbus, OH 43210–1292
[2]Current address: Department of Biology, Massachusetts Institute of Technology, Cambridge, MA 02139

0097–6156/92/0498–0059$06.00/0
© 1992 American Chemical Society

Adams and colleagues (1-5) for information on the electron transport
enzymes hydrogenase, ferredoxin, and a tungsten containing redox protein
from the hyperthermophilic bacterium Pyrococcus furiosus. Also, the
properties of hydrogenase from an extremely thermophilic eubacterium,
Thermotoga maritima was recently described by Adams and coworkers
(6). Most of the work described herein concerns electron transport
components in P. brockii, but some work on the respiratory components
(quinones and cytochromes) from other pertinent thermophilic bacteria is
briefly reviewed for ready comparison to the P. brockii components.

The hyperthermophilic archaebacterium Pyrodictium brockii grows
optimally at 105°C by a form of metabolism known as hydrogen-sulfur
autotrophy, which is characterized by the oxidation of H_2 by S° to produce
ATP and H_2S. The bacterium was isolated from a solfatara field off the
coast of Volcano, Italy by Stetter and his colleagues (7,8). Even though
CO_2 is utilized as the primary if not sole carbon source and H_2 as the
energy source in P. brockii metabolism, it should be noted that growth is
facilitated by the addition of yeast extract to the medium. Nevertheless
the presence of yeast extract does not relieve the requirement for either
H_2 or CO_2 (9,10). The actual sulfur substrate reduced by P. brockii is
probably polysulfide, the product of nucleophilic attack of S-2 on S° rings
(11). Like all hyperthermophiles isolated to date, P. brockii is a strict
anaerobe.
   Hydrogen uptake hydrogenases and their associated electron
transport chains have been well studied in many H_2-oxidizing bacteria, but
H_2 oxidation systems coupled to S° reduction are relatively rare and
poorly-studied. Furthermore, no respiratory-type electron transport
components from any hyperthermophilic bacteria had been identified.
We have identified a quinone and a c-type cytochrome in P. brockii
membranes. Also, we have been able to purify the H_2-uptake
hydrogenase from P. brockii (12,13) and compare it with eubacterial
counterparts. The P. brockii hydrogenase and the H_2-oxidizing respiratory
components (Pihl, T.D., Black, L.K., Schulman, B.A., and Maier, R.J.
1992, Jan. issue, J. Bacteriol.) have some striking similarities to eubacterial
and mesophilic electron transport pathways, and appear not to closely
resemble the H_2-coupled metabolic components of anaerobes.

Hydrogenase Properties

Hydrogenase was purified from P. brockii by use of reactive red as an
affinity agent (13). Reactive red has been used previously to purify some
eubacterial hydrogenases (14,15). The holoenzyme had a molecular mass
of 118,000 ± 19,000 Da, and is composed of one subunit each of 66,000
and 45,000 Da. The large subunit is within the size range for a subunit
commonly observed for hydrogenases, and in fact the P. brockii subunit
reacts with antiserum prepared against the large subunit of hydrogenase

from the aerobic H_2-oxidizing bacterium B. japonicum *(12)*. The artificial electron acceptor specificity and the reversible inactivation effect of O_2 on the P. brockii enzyme is also similar to that of aerobic eubacterial H_2-oxidizers *(12,13)*. The membrane bound enzyme was the only hydrogenase found in the organism and it appears to function in the "H_2-uptake" capacity, linking H_2 oxidation to respiratory components. The membrane bound enzyme did not evolve H_2 even in the presence of low potential electron donors like methyl viologen or benzyl viologen *(12)*.

Metals. The P. brockii hydrogenase contains nickel, and about 8 Fe and 6 acid-labile sulfurs per 118,000 Da protein. Due to the common underestimate in sulfur content, it is likely the enzyme contains two 4Fe-4S clusters. All known NiFeS hydrogenase have at least two subunits, and many have a Ni/Fe/S content similar to that found for P. brockii. The iron content for P. brockii is similar to that for membrane bound uptake hydrogenases from a variety of aerobic eubacteria such as B. japonicum, A. eutrophus and A. vinelandii (see examples Table I). In contrast, the iron content for hydrogenase from the hyperthermophile Pyrococcus furiosus, or the non Ni-containing hydrogenases from anaerobes Clostridium pasteurianum, Megasphaera elsdenii, and D. vulgaris is much greater than for P. brockii (see Table). Other characteristics of the Fe only hydrogenases (such as O_2 sensitivity, *(1)*) also differ from the properties of the P. brockii enzyme *(13)*.

Thermal Stability. Despite the high growth temperature of the bacterium, the purified hydrogenase was still sensitive to thermal destruction. Degraded products could be visualized on SDS gels after prolonged boiling *(13)*, and more than 50% of the methylene-blue dependent activity was eliminated in one h at 98°C. The enzyme was more stable at 90°C, but still half lives of just 1-2 hours were commonly observed *(13)*. The enzyme is much less thermal stable than the hydrogenase *(4)* or ferredoxin *(3)* from the hyperthermophile Pyrococcus furiosus.

Interestingly, the membrane bound form of the enzyme is more tolerant to thermal destruction than the solubilized and purified enzyme (see Figure 1). After incubation for 15 min at 90°C the membrane bound form lost 20% of its initial activity, whereas the purified enzyme lost 70% of its initial activity. Although this result was obtained by a (short) 15 min incubation, this result could be of significance in vivo, as the organism probably has a high enzyme turnover and synthesis rate *(22)*. This affect of the membrane on protection of the enzyme may be due to hydrophobic affects of the membrane lipids on hydrogenase structure. However, the purified enzyme was initially solubilized in Triton X-100 and no steps to remove the detergent were used. Therefore, because of the inherent hydrophobic properties of the detergent itself, a general hydrophobic interaction resulting in thermal protection of the enzyme can

Table 1. Iron and Nickel Content of Selected Hydrogenases

	Nickel*	Iron atoms per molecule	Reference
Bradyrhizobium japonicum	+	~ 6	*16*
Alcaligenes eutrophus (membrane - bd)	+	~ 8	*14,17*
Azotobacter vinelandii	+	~ 6	*18*
Methanosarcina barkeri	+	~ 9	*19*
Pyrodictium brockii	+	~ 8	*13*
Pyrococcus furiosus	+	~ 31	*4*
Clostridium pasteurianum			
Hydrogenase I	-	~ 20	*1*
Hydrogenase II	-	~ 14	*1*
Thermotoga maritima	-	~ 10	*6*
Megasphaera elsdenii	-	~ 16	*20*
Desulfovibrio vulgaris	-	~ 14	*1,21*

*A plus indicates one (or less) mole of nickel per mole of enzyme. Although the $\underline{P}$. brockii enzyme has nickel (based on ^{63}Ni measurements) the amount per mole of enzyme has not been determined.

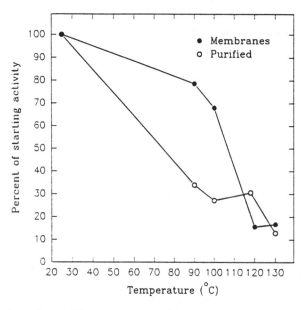

Figure 1. $\underline{P}$. brockii hydrogenase activity as a function of incubation temperature (15 min). Reprinted with permission from ref. 13. Copyright 1991 American Society for Microbiology.

probably be ruled out. The next step needed to address the possible thermal stability provided by membranes is reconstitution of the purified enzyme into defined liposomes.

Km for H$_2$. In addition to playing a role in thermal protection, the membranes appear to play a role in catalytic efficiency. The Km of the membrane bound enzyme is approximately 19 μM whereas the purified enzyme exhibited a Km of about 90 μM (see Figure 2). Although both Km's are within the range commonly seen for bacterial hydrogenase enzymes (either purified or in the membrane) the consistently lower Km we observed for the membrane-bound form points to another potentially helpful facet of the P. brockii membrane in efficient membrane enzyme catalysis. Again, experiments using defined lipids to generate proteoliposomes of known composition will be useful to further investigate this phenomenon.

The P. brockii uptake-type hydrogenase seems to be fairly closely related to the eubacterial uptake-type hydrogenases, in both its physiological and biochemical characteristics. The presence of such a conserved enzyme in an extreme thermophile may be of taxonomic interest, especially given the unsettled phylogenetic status of the thermophiles (23,24). If the membrane itself appears to be the major player in determining thermal stability and catalytic efficiency at high temperatures, it would become clear why the purified P. brockii hydrogenase enzyme itself can be similar in molecular composition to that of the mesophiles. Like the detailed studies on P. furiosus hydrogenase (2,4) our studies provide new enzymological and biochemical data enabling us to compare the diverse ways in which bacteria deal with the activation of the simplest of all substrates, namely, H$_2$. Further research into the differences between the P. brockii hydrogenase and its mesophilic counterparts will hopefully elucidate some of the structural requirements and components that are the basis of thermophily.

Respiratory-type Electron Transport Components

In P. brockii metabolism, electrons from the substrate H$_2$ are presumably passed through an electron transport chain terminating with S$^\circ$ to form H$_2$S. There is no reason to believe that the energy generating mechanisms that occur in hyperthermophiles would differ from the respiration-driven proton translocating system described for other bacteria. The hyperthermophile P. brockii is only a mild acidophile, so that the additional proton-motive force constraints that apply to extreme acidophiles (25) should not be of much consequence. However, the characteristics of the individual respiratory components may differ considerably for hyperthermophiles, and their study should be of value to understand the biochemical basis of biomolecule thermal stability.

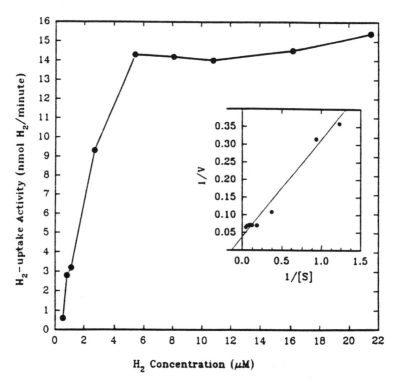

Figure 2. Hydrogen uptake kinetics for the membrane bound P. brockii enzyme. Reprinted with permission from ref. 13. Copyright 1991 American Society for Microbiology.

Quinone. In aerobic H_2-oxidizing bacteria quinones are commonly involved in the sequential H_2-oxidation pathway, but it is not known if quinones are the direct acceptor of electrons from hydrogenase. UV-light is known to inactivate quinones (*26,27*), and UV-light irradiated P. brockii membranes were used to give the first indication that P. brockii utilized a quinone-dependent electron transport chain. H_2-dependent sulfide production was inhibited approximately 80% by the exposure of P. brockii membranes to UV light (Figure 3, Lane B).

Quinones can be added and reconstituted into membranes after dissolving the quinone in absolute ethanol. The addition of the ethanolic quinone-containing solution to an aqueous solution of membranes allows the hydrophobic quinone to insert into the membranes. Hydrogen-dependent sulfide production could be restored to 52% by the addition of 10 μM ubiquinone Q_6 (Lane C). If the known ubiquinone Q_6 was exposed to UV light prior to its addition to the membranes, it was unable to reconstitute hydrogen-dependent sulfide production to the UV irradiated membranes (Figure 3, Lane E). UV light-irradiation did not affect the ability of the membrane bound hydrogen-uptake hydrogenase to oxidize hydrogen when using 200 μM methylene blue as the electron acceptor (data not shown). A quinone with TLC migration properties similar to Q_6 was purified from P. brockii membranes. Subsequent reconstitution experiments with the purified P. brockii quinone showed it was able to restore hydrogen dependent sulfide production to UV irradiated membranes (Figure 3, Lane D). If the ethanolic solution of purified P. brockii quinone was irradiated with UV light, it was unable to restore hydrogen-dependent sulfide production to UV damaged membranes (Figure 3, Lane F). The results implicate the participation of a quinone in the H_2-oxidizing electron transport chain of P. brockii. The use of a quinone as an acceptor of electrons from or near the level of hydrogenase is like that of the aerobic H_2-oxidizing electron transport chains rather than the H_2-metabolizing systems of anaerobes.

The specific structure of the P. brockii quinone is of interest but preliminary NMR analysis has not given definitive information on its structure. The spectra were consistent with a ring structure containing isoprenoid-like units, but the ring structure data did not definitively fit any of a number of quinones, including S-containing quinones from thermophilic acidophiles (*28*) or from H_2-oxidizing thermophiles (*29*). Although P. brockii is a strict anaerobe, its use of a quinone in the H_2-oxidizing chain is similar to the aerobic H_2-oxidizing electron transport chains of bacteria such as B. japonicum and A. vinelandii (*27,30*), as well as other aerobic H_2-utilizing bacteria (*31*). Whether or not the hyperthermophilic archaebacterial H_2-oxidizing chain utilizes menaquinone (present in some eubacterial anaerobes) or ubiquinone (common in eubacterial aerobes) or some other type of quinone must await further physical analyses of the P. brockii quinone.

Cytochrome *c*. A *c*-type cytochrome could be solubilized from P. brockii membranes by use of 0.5% Triton X-100. The Triton X-100- solubilized *c*-type cytochrome had alpha, beta, and gamma peaks at 553, 522, and 421 nm, when reduced with sodium dithionite. The use of Triton X-100 (at 0.5%) was sufficient to solubilize all of the cyt *c* from the membrane (data not shown). Heme staining of SDS-gels P. brockii membranes revealed a single heme-staining component with a molecular mass of 13-14 kDa (see Figure 4). The P. brockii H$_2$-oxidizing chain therefore appears to contain hydrogenase, a quinone and a c-type cytochrome. The presence of a *c*-type cytochrome in P. brockii suggests it may play a key role in hydrogen-sulfur autotrophy. Some strains of *Desulfovibrio* require only cytochrome *c*$_3$, in addition to hydrogenase for the reduction of sulfate (*32*). Thus it is possible that the P. brockii cytochrome *c* also is part of or associated with the terminal S°-reducing protein in electron transport. However, if it is the only cytochrome in the P. brockii membranes (as our results suggest), it probably plays the unusual function (*31*) of a quinol oxidase as well.

Cytochromes from other Thermophiles. Cytochromes of the *c*-type from thermophilic (but not hyperthermophilic) bacteria have been described. The Thermus thermophilus *c*-522 has homology to other *c*-type cytochromes especially in the twenty amino-terminal amino acid residues that contain the heme-linking regions (*33*). Some properties of cytochrome oxidases from the thermophilic bacterium PS3 (*34*) and other thermophilic bacteria (*35,36*) have been described; however these studies do not shed light on the specific forces that enable these cytochromes to withstand higher temperatures.

The cytochrome *c*-552 from Hydrogenobacter thermophilus has structural similarities to other *c*-type cytochromes (*37,38*) even though it can return to its native state after autoclaving at 120°C for 10 min. (*39*). Melting profiles detected by CD spectra in the presence of 1.5 M Gdn-HCl were compared; the T$_{1/2}$ values for the *c*-type cytochromes were: Hydrogenobacter, 90°C; Pseudomonas aeruginosa, 56°C; and horse heart, 54°C. Without Gdn-HCl the Hydrogenobacter enzyme could not be melted. The melting profiles for the mesophile and thermophile are clearly different and these profiles (both with and without guanidinium chloride) are shown in Figure 5. The denaturation events consisted of a single cooperative transition enabling the assignment of clear T$_{1/2}$ melting values. Based on the amino acid sequence of cyt *c*-552 in comparison to cyt *c*-551 of P. aeruginosa, it was hypothesized that additional α-helix structure and electrostatic interaction of the thermophilic cytochrome may relate to its more thermal-stable nature compared to the mesophilic *c*-type cytochromes.

Interestingly, increased thermostability has been introduced into yeast cytochrome *c* by a single amino-acid change. Replacement of wild type Asn-52 with Ile-52 results in a mutant form of the cytochrome that

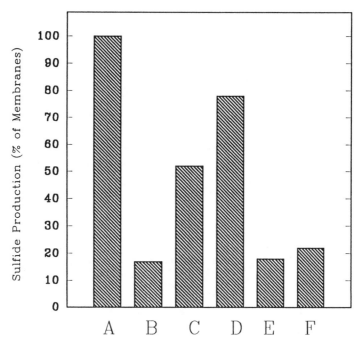

Figure 3. Reconstitution of H_2-dependent sulfide production in UV-light treated membranes of P. brockii.
Lanes A (untreated membranes), B (UV-light irradiated), C (addition of ubiquinone Q_6), D (addition of P. brockii TLC-purified quinone), E (addition of UV-light treated Q_6), F (addition of UV-light treated P. brockii quinone).

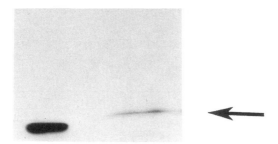

Figure 4. Heme staining of SDS-gels of (left lane) horse heart cytochrome *c*, and (right lane, arrow) P. brockii membranes. Reprinted with permission from ref. 46. Copyright 1992 American Society for Microbiology.

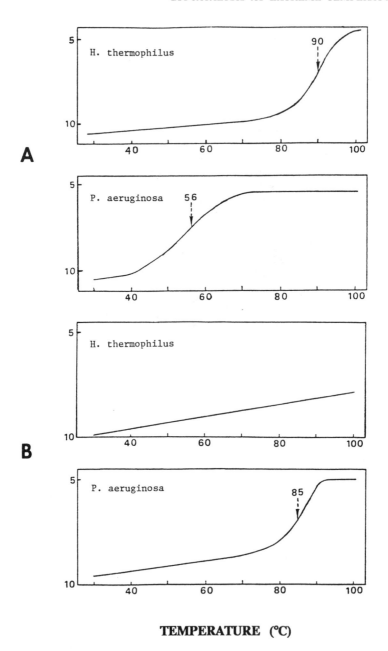

TEMPERATURE (°C)

Figure 5. Thermal denaturation curves of H. thermophilus and P. aeruginosa c-type cytochromes in the presence of (A) and absence (B) of 1.5 M Gdn-HCl. Adapted from ref. 39. The x-axis is $-[\Theta]_{222}$ x 10^{-3} (see ref. 39).

has twice the free energy of denaturation of the wild type form (*40,41*). Studies with this and other single site mutants suggested hydrophobic interactions are the main factor responsible for enhancing thermostability of the Ile-52 mutant.

Sulfur Reduction. Bacterial dissimilatory sulfur metabolism is complicated in that a variety of sulfuric compounds can serve as substrates. These include sulfate, sulfite, tetrathionate, trithionate, thiosulfate, elemental sulfur and sulfide (see ref. *22*). Hydrogen is an important electron donor in both sulfate and sulfur reduction, and P. brockii couples H_2 oxidation to elemental sulfur reduction. However, polysulfide is probably an important terminal electron acceptor for P. brockii (*11*). In contrast to sulfate reduction (such as in Desulfovibrio spp.) there is unfortunately little information on use of other sulfur sources in dissimilatory S metabolism. Elemental sulfur reduction catalyzed by a sulfur reductase has been characterized as a c_3-type cytochrome from several sulfate-reducing bacteria (*32*). Interestingly, the H_2-dependent reduction of sulfur could be catalyzed in vitro with only the purified cytochrome and hydrogenase. This implies that the cytochrome can serve as a hydrogenase oxidase. For P. brockii at least one other component (a quinone) is in the H_2/S electron transport chain. Whether the P. brockii c-type cytochrome can by itself fulfill the sulfur reductase role is not yet known.

A sulfur reductase from Wolinella succinogenes has also been purified (*44*). It does not contain heme, and it could couple electron transport from formate to sulfur when reconstituted into liposomes. Also, in addition to hydrogenase, a b-type cytochrome, and an iron sulfur protein associated with sulfur reduction, two quinones were identified in the sulfur-reducing bacterium Spirillum 5175 (*42,43*). One of the quinones was identified as menaquinone MK6 and the other was at too low a concentration to allow its characterization (*43*). The electron carrying sequence of electron transport components in Spirillum 5175 along with characterization of their mid-point potentials will undoubtedly provide us with a novel H_2-oxidizing electron transport pathway for comparison to other S^o-reducers. For studying the bioenergetic coupling abilities of P. brockii in detail, the hydrogenase, quinone, c-type cytochrome, and sulfur reductase will all have to be incorporated into liposomes. To what extent E. coli, soybean, or hyperthermophilic organism lipids will suffice for such studies will be interesting.

Conclusions

We propose the following model for electron transport in P. brockii as being the minimum that is consistent with our data (Figure 6). First, hydrogen is oxidized by the membrane bound hydrogen-uptake hydrogenase. The electrons generated by this oxidation are then used to

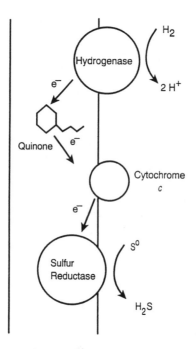

Figure 6. Proposed H_2-oxidizing S°-reducing electron transport pathway in <u>P</u>. <u>brockii</u> membranes.

reduce a quinone, which in turn is oxidized by a c-type cytochrome. The membrane bound character of the hydrogenase and the membrane association of the cytochrome may reflect the requirement of these proteins for interacting with the hydrophobic quinone. The model is not rigorously proven nor necessarily complete as yet. For example, we have not yet identified the terminal sulfur reductase, and reconstitution experiments will be needed to determine if the quinone is the direct electron acceptor from hydrogenase. Also, there may be other components such as a ferredoxin or flavodoxin, and although the c-type cytochrome may fill the sulfur reductase role, we have not attempted to assign this function to the c-type cytochrome. We also cannot yet propose a "sidedness" to the components of this electron transport chain. The production of scalar protons are of importance in the energy-generating metabolism of several H_2-utilizing bacteria, particularly <u>Desulfovibrio</u>, however, for <u>P</u>. <u>brockii</u> there is currently no information available to allow a suggestion that production of scalar protons occurs. It should be noted that the energy couple for H_2 to S^o is only 144 mV (*45*). This limited energy potential might constrain the number of components permitted to efficiently capture energy in this presumably primitive electron transport chain.

Literature Cited

1. Adams, M.W.W. *Biochim. Biophys. Acta.* **1990**, *1020*, 115-145.
2. Adams, M.W.W. *FEMS Microbiol. Rev.* **1990**, *75*, 219-238.
3. Aono, S., Bryant, F.O., Adams, M.W.W. *J. Bacteriol.* **1989**, *171*, 3433-3439.
4. Bryant, F.O., Adams, M.W.W. *J. Biol. Chem.* **1989**, *264*, 5070-5079.
5. Conover, R.C., Kowal, A.T., Fu, W., Park, J.-B., Aono, S., Adams, M.W.W., and Johnson, M.K. *J. Biol. Chem.* **1990**, *265*, 8533-8541.
6. Juszczak, A., Aono, S., Adams, M.W.W. *J. Biol. Chem.* **1991**, *266*, 13835-13841.
7. Stetter, K.O. *Nature* (London) **1982**, *300*, 258-760.
8. Stetter, K.O., H. Köning, and E. Stackebrandt. *Syst. Appl. Microbiol.* **1983**, *4*, 535-551.
9. Parameswaran, A.K., Provan, C.N., Sturm, F.J., and Kelly, R.M. *Appl. Environ. Microbiol.* **1987**, *55*, 1690-1693.
10. Parameswaran, A.K., Schicho, R.N., Soisson, J.P., and Kelly, R.M. *Biotechnol. Bioeng.* **1988**, *32*, 438-443.
11. Blumenthals, I.I., Itoh, M., Olson, G.J., and Kelly, R.M. *Appl. Environ. Microbiol.* **1990**, *56*, 1255-1262.
12. Pihl, T.D., Schicho, R.N., Kelly, R.M., and Maier, R.J. *Proc. Natl. Acad. Sci. USA* **1989**, *86*, 138-141.
13. Pihl, T.D., Maier, R.J. *J. Bacteriol.* **1991**, *173*, 1839-1844.
14. Schneider, K., Pinkwart, M., and Jochin, K. *Biochem. J.* **1983**, *213*, 391-398.
15. Stults, L.W., Moshiri, F., and Maier, R.J. *J. Bacteriol.* **1986**, *166*, 795-800.

16. Arp, D.J. *Arch. Biochem. Biophys.* **1985**, *237*, 504-512.
17. Schneider, K., Patil, D.S., Cammack, R. *Biochim. Biophys. Acta.* **1983**, *784*, 353-361.
18. Seefeldt, L.C., Arp, D.J. *Biochimie.* **1986**, *68*, 25-34.
19. Fauque, G., Teixeira, M., Moura, P.A., Lespinat, A.V., Xavier, A.V., DerVartanian, D.V., Peck, H.D., LeGall, J., Moura, J.G. *Eur. J. Biochem.* **1984**, *142*, 21-28.
20. Filipiak, M., Hagen, W.R., Veeger, C. *Eur. J. Biochem.* **1989**, *185*, 547-553.
21. Hagen, W.R., van Berkel-Arts, A., Krüse-Wolters, K.M., Voordouw, G., Veeger, C. *FEBS Microbiol. Lett.* **1986**, *203*, 59-63.
22. Pihl, T.D., Schicho, R.N., Black, L.K., Schulman, B.A., Maier, R.J., and Kelly, R.M. *Biotech. Genetic Eng. Rev.* **1990**, *8*, 345-377.
23. Woese, C.R. *Microbiol. Rev.* **1987**, *51*, 221-271.
24. Lake, J.A. *Nature* (London) **1990**, *343*, 418-419.
25. Krulwich, T.A., Ivey, D.M. In *Bacterial Energetics*; Krulwich, T.A., Ed.; Academic Press, Inc., San Diego, CA., 1990, pp. 417-447.
26. Erickson, S.K., and Parker, G.L. *Biochim. Biophys. Acta* **1969**, *180*, 56-62.
27. Wong, T.-Y., and Maier, R.J. *J. Bacteriol.* **1984**, *159*, 348-352.
28. DeRosa, M., DeRosa, S., Gambacorta, A., Minale, L., Thompson, R.H., and Worthington, R.D. *J. Chem. Soc. Perkin. Trans.* **1977**, *1*, 653-657.
29. Ishii, M., Kawasumi, T., Igarashi, Y., Kodama, T., and Mionda, Y. *J. Bacteriol.* **1987**, *196*, 2380-2384.
30. O'Brian, M.R., and Maier, R.J. *J. Bacteriol.* **1985**, *161*, 775-777.
31. Anraku, Y. *Ann. Rev. Biochem.* **1988**, *57*, 101-132.
32. Fauque, G., Herve, D, and LeGall, J. *Arch. Microbiol.* **1979**, *121*, p.261.
33. Titani, K., Ericsson, L.H., Hon-nami, K., Miyazawa, T. *Biochem. Biophys. Res. Commun.* **1985**, *128*, 781-787.
34. Baines, B.S., Poole, R.K. In *Microbial Gas Production*; Poole, R.K. and Dow, C.D., Eds.; Society for General Microbiol., Academic Press, New York, 1985, pp. 63-74.
35. Hon-nami, K., and Oshima, T. *Biochem. Biophys. Res. Commun.* **1980**, *92*, 1023-1029.
36. Hon-nami, K., and Oshima, T. *Biochem.* **1984**, *23*, 454-460.
37. Ishii, M., Itoh, S., Kawasaki, H., Igarashi, Y., Kodama, T. *Agric. Biol. Chem.* **1987**, *51*, 1825-1831.
38. Sanbongi, Y., M. Ishii, M., Igarashi, Y., and Kodama, T. *J. Bacteriol.* **1989**, *171*, p.65.
39. Sanbongi, Y., Igarashi, Y., and Kodama, Y. *Biochem.* **1989**, *28*, p. 9575.
40. Das, G., Hickey, D.R., McLendon, D., McLendon, G, and Sherman, F. *Proceed. Nat'l Acad. Sci.* **1989**, *86*, 496-499.
41. Hickey, D.R., Berghuis, A.M., Lafond, G., Jaeger, J.A., Cardillo, T.S., McLendon, D., Das, G., Sherman, F., Brayer, G.D., and McLendon, G. *J. Biol. Chem.* **1991**, *266*, 11686-11694.

42. Zöphel, A., Kennedy, M.C., Beinert, H., and Kroneck, P.M.H. *Arch. Microbiol.* **1988**, *150*, p.72.
43. Zöphel, A., Kennedy, M.C., Beinert, H., and Kroneck, P.M.H. *Eur. J. Biochem.* **1991**, *195*, p.849.
44. Schröder, J., Kröger, A., and Macy, J.M. *Arch. Microbiol.* **1988**, *149*, p.572.
45. Thauer, R.K., Jungermann, K., and Decker, K. *Microbiol. Rev.* **1977**, *41*, 100-180.
46. Pihl, T.D, Black, L.K., Schulman, B.A., and Maier, R.J. *J. Bacteriol.* **1992**, *174*, 137-143.

RECEIVED January 15, 1992

Chapter 6

Key Enzymes in the Primary Nitrogen Metabolism of a Hyperthermophile

Frank T. Robb[1], Yaeko Masuchi[1], Jae-Bum Park[2], and Michael W. W. Adams[2]

[1]Center for Marine Biotechnology, University of Maryland, Baltimore, MD 21202
[2]Department of Biochemistry and Center for Metalloenzyme Studies, University of Georgia, Athens, GA 30602

This study focuses on the primary pathways of nitrogen metabolism in the hyperthermophile *Pyrococcus furiosus*. Two key enzymes, glutamine synthetase (GS) (EC 6.3.1.2) and glutamate dehydrogenase (GDH) (EC 1.4.1.3), have been characterized. The addition of pyruvate and maltose to continuous cultures of *P. furiosus* resulted in 10-fold and 40-fold decreases in GDH activity, respectively. GS and protease levels were slightly decreased by these growth conditions, except for the addition of pyruvate, which increased GS activity slightly. These data support our conclusion from kinetic analysis that GDH is adapted for a catabolic function, and forms a pathway for utilization of amino acids. GS was found to have a $M_r = 330,000$ kD. In the absence of magnesium ions, dissociation of the GS was observed. Subunits of $M_r=58,000$ could be reconstituted to form active enzyme complexes in the presence of 5mM Mg^{++}. GDH was found to be a major activity (>1%) of crude extracts of *P. furiosus*. GS and GDH from *P. furiosus* represent the most thermostable versions of these enzymes described to date.

The discovery of microorganisms that grow optimally near or above 100°C (*1, 2*) raises the question of how metabolic processes occur at extremely high temperatures. Since these organisms are presumably unable to control their internal temperature, metabolic activity must depend upon the intrinsic thermostability and catalytic properties of enzymes. The organism used in this study, *Pyrococcus furiosus*, was first isolated from shallow marine solfataras and was described in 1986 by Fiala and Stetter (*3*). It grows optimally at a temperature of 100°C. It belongs to a diverse group of microorganisms that grow optimally near 100°C, all of which are members of the Archaea, formerly known as the Archaebacteria (*4*).

0097–6156/92/0498–0074$06.00/0
© 1992 American Chemical Society

P. furiosus is a strictly anaerobic heterotroph that grows by the fermentation of peptides and certain carbohydrates to produce H_2 (or hydrogen sulfide if inorganic sulfur, $S°$, is present), organic acids and CO_2 (*2*). Interestingly, optimal growth requires peptides as the N source rather than ammonia or amino acids (*2,5*). Moreover, the optimal growth rate and yield during fermentation is only achieved upon addition of tungsten to the growth medium (*5-8*). The purification of a novel tungsten-containing iron-sulfur oxidoreductase that catalyzes the oxidation of aldehydes has been described recently (*9*), and this may indicate a unique aspect of hyperthermophile metabolism leading to a dependence on tungsten, an element that is rarely encountered in biological systems.

The question arises as to the metabolic relationship between peptide utilization and the metabolism of carbon and nitrogen in this organism.   Peptides can be used as the sole supply of carbon and nitrogen by *P. furiosus*.   Since amino acid utilization must be an important metabolic process, we have studied the enzymes that constitute the central nitrogen metabolism in *P. furiosus,* namely glutamine synthetase (GS) and glutamate dehydrogenase (GDH). GS is a ubiquitous enzyme responsible for the synthesis of glutamine from glutamate and ammonia. Glutamine is the common precursor in the formation of the majority of the nitrogen containing compounds in most organisms. Because GS performs this crucial biosynthetic function it is important to consider the regulation and catalytic properties of this enzyme from a hyperthermophile. GDH can function either in a biosynthetic role, in the formation of glutamate, or in the catabolism of glutamate and the generation of 2-oxo-glutarate, an important intermediate in carbon and energy metabolism. It is thus in a pivotal position between the major pathways of carbon utilization and anabolic nitrogen metabolism.   The purification of GDH from *P. furiosus* was recently reported (*10*). In this study we also describe the apparant regulation in *P. furiosus* of both GS and GDH by carbon and nitrogen growth substrates, and the purification, thermostability and subunit structure of GS.

Methods

Bacterial strain and cultivation. *P. furiosus* (DSM 3638) was grown as closed static cultures in synthetic sea water supplemented with a vitamin mixture, $FeCl_3$ (25 μM), elemental sulfur (5 g/L, w/v) and Na_2WO_4 (10 μM) as previously described (*8,9*). The synthetic sea water medium (13), consisting of NaCl (24 g/L), Na_2SO4 (4g/L), KCl (0.7 g/L), Na HCO_3 (0.2 g/L), KBr (0.1 g/L), H_3BO_3 (30 mg/L), $MgCl_2.6$ H_2O (10.8 g/L), $CaCl_2.2H_2O$ (1.5 g/L), $SrCl_2.6H_2O$, (25 mg/L), sodium resazurin (0.2 mg/L), was supplemented tryptone (1 g/L: low tryptone medium). Cells stored at 4°C in this medium remained viable for at least a year. Large scale growth was carried out at 88°C in the absence of sulfur but with titanium (III) nitrilotriacetate (final concentration, 30 μM) as a reductant in a 500 liter stainless steel fermentor, as previously described (5, 8, 9). Cultures were sparged with argon at a rate of 7.5 L/min. For physiological studies, the continuous culture system described by Brown and Kelly (11) was used. It was maintained at 98°C and sparged with prepurified N_2 at a rate of 50 ml/min to achieve anaerobic conditions, as judged by the oxidation state of the resazurin. Pyruvate (12), maltose (2) and glutamate were added as additional carbon sources at 4 g/L where indicated. For the experiment shown in Figure 1, the sea water medium was supplemented with 5 g/L tryptone.

Preparation of Extracts. *P. furiosus* cells grown under optimal conditions in a 500 L fermentor were harvested and lysed as described previously (6, 7). Continuous cultures maintained at 99°C were used to obtain cells grown under different nutritional conditions. They were resuspended at 20% (w/v) in imidazole buffer, pH 7.15, containing 10 mM dithiothreitol and 5 mM Mg Cl_2, and were lysed by two passages through the French pressure cell at 20,000 psi. A cell-free extract was obtained by centrifugation at 48,000xg for 30 mins.

Enzyme Purification. GS was partially purified using 600 g of cells (wet weight) as starting material. The initial purification procedure was the same as for *P. furiosus* ferredoxin, up to and including the first Q-Sepharose column (6). This and all subsequent columns were controlled by a Pharmacia FPLC system. Fractions containing the GS were combined, and were applied to a column (5 x 30 cm) of DEAE Sephadex, equilibrated with 50 mM Tris/HCl buffer, pH 8.0. Adsorbed proteins were eluted with a gradient (2000 ml) from 0 to 1.0 M NaCl. The GS activity coeluted with α-glucosidase at a position corresponding to 0.29 M NaCl. The pooled fractions containing GS activity were concentrated to about 20 ml using an Amicon ultrafiltration cell fitted with a PM-30 membrane, and were applied to a column (5 x 95 cm) of Sepharose CL-6B, equilibrated with 50 mM Tris/HCl buffer, pH 8.0, containing 0.2 M NaCl, and eluted at 3 ml/min. GS eluted from the Sepharose CL-6B column prior to the α-glucosidase, at a position that corresponded to an M_r of about 300,000. This material was used to study the dissociation of GS by Superose 12 chromatography (see below).

Enzyme Assays and Other Methods. The reaction used to detect GS activity is as follows:

$$\text{Glutamine} + NH_2OH + ADP \quad \xrightarrow[\text{Arsenate}]{Mn^{++}} \quad \text{γ-Glutamyl hydroxamate} + AMP$$

The γ-glutamyl transferase assay mixture contained 18 mM hydroxylamine hydrochloride, 0.27 mM $MnCl_2$, 25 mM disodium hydrogen arsenate, 0.36 mM ADP, in 135mM imidazole HCl buffer, pH 7.15. To 0.4 ml of assay mixture, sample and water were added to a final volume of 0.45 ml, which was preincubated in a temperature controlled aluminium block at 85°C for 5 min before the reaction was initiated by the addition of 50 μL of 0.2 M glutamine. The assay was linear for 10 mins at 85⁰C. For controls and blanks, ADP was omitted.

To terminate the reaction and generate color, 1 ml of a solution containing $FeCl_3.6H_2O$ (55 g/L), trichloroacetic acid (20 g/L) and concentrated HCl (21 ml/L) was added. The mixture was chilled briefly on ice, vortexed and centrifuged to remove any precipitate before reading the OD at 540 nm. In this assay, 1 μmol of γ-glutamyl hydroxamate gives 0.533 OD units at 540 nm; one unit of GS was defined as producing 1 μmol of γ-glutamyl hydroxamate per min at 85°C.

GDH activity was measured by the glutamate-dependent reduction of NADP at 85°C (*10*). The reaction catalyzed by glutamate dehydrogenase is shown below:

Glutamate + NADP   <===>   2-ketoglutarate + NADPH + NH_3

The progress of the forward and reverse reactions were followed by the absorbance of NADPH at 340 nm, using a thermostatted multiple cuvette carrier in a Beckman DU50 spectrophotometer. One unit of GDH activity is defined as 1 μmole of NADPH formed per min. The α-glucosidase activity was assayed using the chromogenic substrate p-nitrophenyl α-D-glucopyranoside, at 85°C in a Spectronic 20 specrophotometer equipped with a temperature regulated cuvette holder (*13*). The absorbance at 405 nm was used to detect hydrolysis of the substrate and the release of p-nitrophenol in sodium phosphate buffer, pH 5.5.

Protease activity was assayed using azocasein as a substrate at 98°C, in 100 mM $NaPO_4$ buffer, pH 6.8 (*14, 15*). Specific activity is defined as the change in absorbance at 440 nm per hr per mg protein. Protein concentrations were estimated by the colorimetric Lowry method using bovine serum albumin as the standard (*16*). The molecular weights of GS and GDH were determined by gel filtration using two connected columns (HR 10/30) of Superose 12 operated by a Pharmacia FPLC system. This was calibrated with aldolase (158,000), bovine serum albumin (67,000), ovalbumin (45,000), carbonic anhydrase (29,000), myoglobin (16,900), and cytochrome c (12,300), using 50mM Tris/HCl buffer, pH 8.0, containing 0.2 M NaCl.

Results

Purification and characterization of GDH. GDH was purified from *P. furiosus* to electrophoretic homogeneity (*10*). The overall purification was 75-fold with a yield of activity of 36%. The enzyme has a molecular weight of 270,000 and is a hexamer of identical subunits. The half-life of the pure enzyme at 100°C was approximately 10 hr, making it the most thermostable dehydrogenase yet purified. *P. furiosus* GDH showed a high K_m for ammonia (9.5 mM), suggesting that the physiological reaction is in the catabolic direction generating 2-oxoglutarate. It displayed a preference for NADP over NAD as cofactor at all temperatures, but this was less pronounced at the optimum temperature for activity, which was 95°C.

Purification and characterization of GS.   Table I summarizes the
results of the partial purification of GS.   The enzyme was relatively
stable at room temperature during several purification steps.
However, the passage of the enzyme over a gel permeation column led
to a precipitous decline in activity, as shown in Table I.   Some activity
was recovered by the overnight dialysis of fractions of low molecular
weight (<100,000) against buffer solutions containing Mn^{++} or Mg^{++}
ions.

Table I. Purification of *P. furiosus* glutamine synthetase (GS)

Step	Activity	Specific Actvitiy	Yield	Purification
	(units)	(units/mg)	(%)	(-fold)
Cell-free extract	4.14	0.011	100	1
Q Sepharose	7.36	0.45	177	42
DEAE Sephadex	7.80	1.63	163	147
Sepharose Cl-6B	0.07	0.11	1.6	10

Figure 1 shows the results of successive experiments on
Superose 12 gel permeation columns using buffer with and without
5mM $MgCl_2$.   GS is stabilized as high molecular weight aggregates of
about M_r 330,000 in the presence of Mg^{++} but is dissociated into subunits
in the absence of divalent cations.   The preincubation at 85°C of
material from fractions of around M_r 62,000 resulted in the recovery of
a small amount of activity in an asymmetrical peak, as shown in
Figure 2.   The presence of α-glucosidase activity in these experiments
provided a convenient internal molecular weight marker, with M_r
=146,000 (*13*).

GS therefore appears to have an apparent molecular weight of
330,000 ± 15,000 estimated by gel filtration in the presence of Mg^{++} and a
subunit molecular weight of 53,000 ± 3000.   These results suggest that it
is a hexameric species.

Figure 2 shows the effect of temperature on the reaction rate of a
GS preparation purified 147-fold.   The enzyme displays an optimum for
the transferase reaction of about 85°C.   The same preparation also
showed no loss of activity when incubated aerobically at 20 or at 4°C
over a period of at least six months.   This preparation of GS was about
50% pure, as judged by polyacrylamide gel electrophoresis.   It is not
known to what extent stability is affected by contaminating proteins.
Figure 3 shows the marked resistance to thermal denaturation that is
evident in partially purified fractions of *P. furiosus* GS.   The
thermostability of the enzyme increased when incubations were
performed using concentrated fractions of partially pure enzyme.
Thus, the half life of a crude extract was 0.65 hours at 100°C, compared

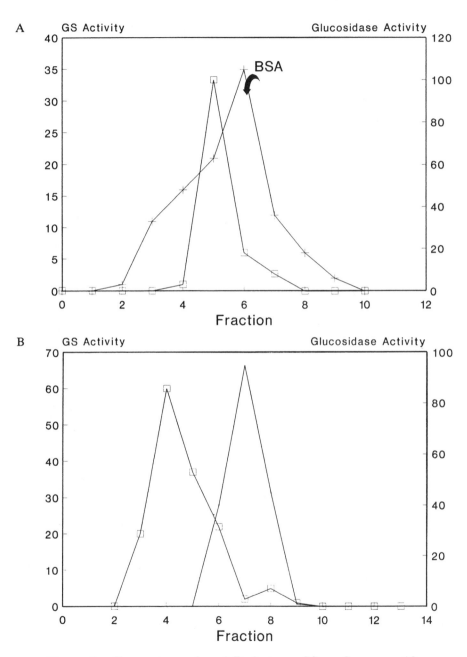

Figure 1. Chromatography of *P. furiosus* GS on Superose 12.
A: the column was equilibrated with 50 mM Tris/HCl buffer,
pH 8.0, containing 0.2 M NaCl. B: as in A except the buffer
contained $MgCl_2$ (5 mM). The internal standard was α-
glucosidase that copurified with GS.

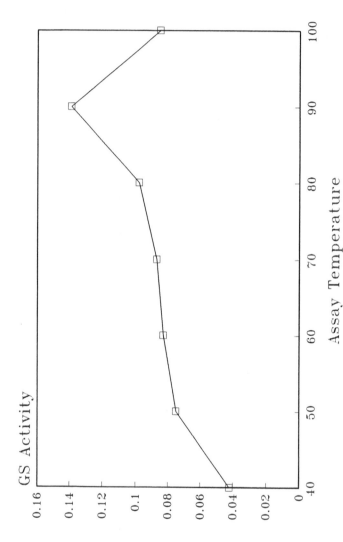

Figure 2. Effect of temperature on the activity of GS from *P. furiosus*. The activity was determined by the glutamyl transferase assay at the indicated temperature.

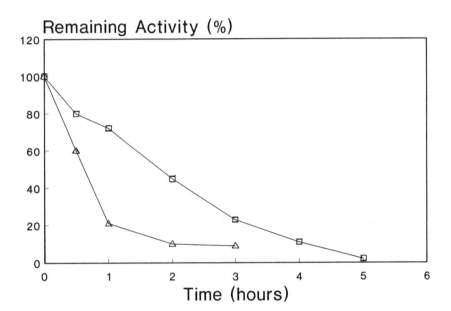

Figure 3. Thermal stability of GS from *P. furiosus.* Partially purified GS, from the peak fraction of the DEAE Sephadex step in Table 2, was heated at the indicated temperature in imidazole buffer, pH 7.15, containing 10 mM dithiothreitol.

with a half life of 1.8 hours when the enzyme was purified 20-fold and incubated at 100°C under otherwise identical conditions.

Regulation of GS, GDH and Protease activities. Table II shows GS, GDH and total protease activities in cell extracts of *P. furiosus* grown in continuous cultures that were limited by low concentrations of tryptone.

Table II. Effect of growth conditions on the expression of GDH and GS in continuous cultures of *P. furiosus*

Growth Condition	Specific Activity (Fold Increase/Decrease)[a]		
	GS	GD	Protease
Low tryptone	0.26 (1.0)	4.13 (1.0)	44 (1.0)
L-glutamate	0.11 (0.4)	0.42 (0.1)	32 (0.7)
L-glutamate + maltose	0.13 (0.5)	0.010 (0.002)	15 (0.3)
L-glutamate + pyruvate	0.37 (1.4)	0.047 (0.011	28 (0.6)

[a]Specific activities are expressed in units/mg and are means of at least two determinations. The cells were grown in low tryptone medium in continuous culture as described as described in Material and Methods.

When the growth limitation was relieved by glutamate, a 10-fold decrease of GDH activity occurred while the GS and protease levels were slightly lowered. Addition of pyruvate and maltose to glutamate supplemented cultures resulted in 40-fold and 10-fold decreases of GDH activity, respectively. Maltose addition had no effect on the GS level, but decreased protease activity 3-fold. The addition of pyruvate resulted in a 50% increase in GS activity.

Discussion

P. furiosus is able to grow with peptides as the sole N and C source, therefore, it is likely that GS and GDH are two key enzymes in the primary metabolism of this organism. We were unable to detect the presence of glutamate synthase in cell extracts. This is usually an important enzyme in the nitrogen metabolism of most organisms. A prior study of the kinetics of *P. furiosus* GDH provided evidence that its most likely function is to degrade glutamate (*10*). This is consistent with the data presented here, in which the effects of the addition of alternative growth substrates on the activities of GDH and GS were investigated. The addition of both pyruvate and maltose decreased GDH activity, although very little variation occurred in the levels of GS. This supports our contention that GDH is a key enzyme in a pathway for the utilization of glutamate as a carbon and energy source. The addition of maltose presumably stimulates an alternative, glycolytic pathway that supplies the cell's requirements for carbon and energy. Pyruvate, which can be used by *P. furiosus* as the sole carbon source for growth (*15*), also partially or completely supplants amino acids in this regard.

In contrast to GDH of *P. furiosus*, GS is regulated over a narrow range, and moreover is apparently expressed at a relatively low level, judging by its low specific activity in crude extracts. This result may be due in part to the nature of the GS assay, which contains several relatively heat labile components, such as ADP and glutamine. The assay temperature of 85°C was chosen because it is the temperature optimum for enzyme activity and was the highest temperature at which linear reaction rates were recorded for periods of more than 10 min. It is possible that unknown mechanisms provide for enhanced activity of GS *in vivo* and that the optimum temperature for the GS reaction is considerably higher than 85°C. The optimum reaction temperature for GDH activity is about 97°C (*10*), although the experiments shown here are based on a reaction temperature of 85°C because of the large correction factors that are necessary to compensate for the thermal degradation of the cofactor, NADPH. This has a half-life of 2.3 min at 100^0C under the conditions used to assay GDH (*10*).

Several remarkably thermostable enzymes have been purified from cell-free extracts of *P. furiosus*. These proteins fall into two groups: first, hydrolytic enzymes involved in the catabolism of proteins and carbohydrates, such as α-glucosidase (*13*) and proteases (*14, 15*), and second, enzymes involved in the conversion of C3 compounds to CO_2 and molecular H_2. The latter category includes a unique hydrogenase, which is, in contrast to all other hydrogenases studied, adapted for evolution of H_2 rather than H_2 uptake (*5, 6*). In addition, extremely thermostable ferredoxin (*7*) and glyceraldehyde (*8,9*) and pyruvate oxidoreductases (Blamey, J and Adams, M.W.W, unpublished data) have been purified from from this organism. *P. furiosus* GS and GDH are, therefore, the first examples of enzymes of nitrogen metabolism that have been characterized from a hyperthermophilic species.

The GS described here is interesting because of its relatively low molecular weight of 330,000. For example, GS from *E.coli*, whose three dimensional structure has been determined, has an M_r value of 620,000 (17) and contains twelve identical subunits of M_r 53,000. In general, bacterial GS enzymes are dodecameric structures composed of two hexameric rings in a double doughnut arrangement. However, a recent study indicated that GS from the obligate anaerobic bacterium, *Bacteroides fragilis*, has an M_r value of 450,000 and is a hexamer similar to GS from *P. furiosus* (18). GS has also been partly characterized from two methanogens, *Methanococcus voltae* (19) and *Methanobacterium ivanovi* (20), and from the aerobic extreme thermophile, *Sulfolobus sulfataricus* (21), However, the subunit structures of the GSs described in these studies were not reported. Partially purified GS from *P. furiosus* had a half life of about 2 hr at 100°C making it the most thermostable of the GS preparations described to date (cf *17-20*).

Our prior work on GDH from *P. furiosus* indicated that it, too is a hexameric enzyme (*10*). The GDHs from *P. furiosus* (*10*) and *S.*

solfataricus (*23*) and the *P. woesei* glyceraldehyde-3-phosphate de-hydrogenase (*24*) all recognise both NAD and NADP. Based on this observation, it is possible that indiscriminate use of both nicotinamide cofactors may be a general property of dehydrogenases from extremely thermophilic Archaea. In general, prokaryote GDH enzymes are specific for either NAD in the case of the catabolic enzymes, or NADP in the case of the biosynthetic ones, whereas the eucaryotic enzymes generally can utilize both cofactors (*25*). Yeast is an exception to this rule, and has been found to have two distinct GDH enzymes with kinetic properties that allow them each to function only in one physiological role (*25, 26*). The ability to use either NAD or NADP cofactors appears to be a Eukaryotic feature of the Archaeal dehydrogenases that can be added to many other such features, for example, introns, yeast-like RNA polymerases and promoter sequences.

The phylogenetic position of the extremely thermophilic Archaea, the Domain to which *P. furiosus* has been assigned, is more closely aligned with the Eukarya than the Bacteria (*4*). Highly conserved enzymes such as GS and GDH are therefore of great interest since extreme thermophiles like *P. furiosus* are amongst the most deeply rooted organisms on the evolutionary hierarchy devised from ribosomal RNA sequencing (*4*). Both GS and GDH are potentially extremely useful models for the evolution of proteins from these prototypical organisms. Our ongoing study will provide information about the evolution of these enzymes, and the mechanisms of their extreme thermostability.

Acknowledgments

This research was supported by grants from the Office of Naval Research (N00014-90-J-1823 to FTR and N00014-90-J-1894 to M. W. W. A), the National Science Foundation (BCS 9011583), and by a National Science Foundation Training Group Award to the Center of Metalloenzyme Studies of the University of Georgia (DIR 9014281). We thank Ilse Blumentals and Christina Chiarantano for help with continuous culture techniques, and Robert Kelly and Allen R. Place for useful advice. This is Contribution Number 182 from the Center of Marine Biotechnology.

Literature Cited

1. Stetter, K.O. *Nature* **1982,** *300,* 258-260.
2. Stetter, K.O.; Fiala; G. Huber, R. ; Segerer A. *FEMS Microbiol. Rev.* **1990,** *75,* 117-124.
3. Fiala, G.;Stetter, K.O. *Arch. Microbiol.* **1986,***145,* 56-61.
4. Woese, C.R.; Kandler, O.; Wheelis,M.L. *Proc. Natl. Acad. Sci. USA* . **1990,** *87,* 4576-4580.
5. Adams, M.W.W. *FEMS Microbiol. Rev.* **1990,** *75,* 125-157.
6. Bryant, F.O.; Adams, M.W.W. *J. Biol. Chem.* **1990,***.265,* 11508-11516.
7. Aono, S.; Bryant, F.O.;Adams, M.W.W.*J. Bacteriol,* **1989,** *171,* 3433-3439.

8. Mukund, S.; Adams, M.W.W.*J. Biol. Chem.,*1990. *265,* 11508-11515.
9. Mukund, S.; Adams, M.W.W. *J. Biol. Chem.,* 1991. *266,* 14208-14216.
10. Robb, F. T.; Park, J-P, Adams, M. W. W. *Biochim. Biophys. Acta.* in the press.
11. Brown, S. H.; Kelly, R. M. *App. Env. Micro.* 1989, *55,* 2086-2088
12. Costantino, H.R.; Brown, S.H.; Kelly, R.M. *J. Bacteriol.* 1990,.*172,* 3654-3660.
13. Blumentals, I.I.; Robinson, A.S.; Kelly, R.M. *Appl. Env. Micro.* 1990, *56,* 1992-1998.
14. Eggen, R.; Geerling, A.;Watts, J.; de Vos, W.M. *FEMS Microb. Lett.* 1990, *71,* 17-20.
15. Schafer, T; Schonheit P. *Arch. Microbiol.* 1991, *155,* 366-377.
16. Lowry, O.H.; Rosebrough, N. J.; Farr,A. L.; Randall, R. J. *J. Biol. Chem.,*1951, *193,* 265-275.
17. Almassy, R.J.; Janson, C.A.; Hamlin R.; Xuong N-H.; Eisenberg, D.*Nature* 1986. *323,* 304-309
18. Hill, R. T.; Parker, J.R., Goodman, H, Jones, D. T., Woods, D. R. *J. Gen Microbiol.* 1989,*135,* 3271-3279.
19. Possot, O.; Sibold, L., Aubert, J.-P. *Res. Microbiol..* 1989, *140,* 355-371.
20. Bhatnagar L.; Zeikus J.G.; Aubert J-P. *J. Bacteriol.* 1986, *165,* 638-643.
21. Sanangelantoni.; A. M.; Barbarini, D; Di Pasquale, G; Cammarano, P.;Tiloni, O. *Mol. Gen. Genet.* 1990, *221,* 187-194.
22. Consalvi, T.; Chiaraluce, R.; Politi, L.; Gambacorta, A; De Rosa , M.; Scandura, R. *Eur. J. Biochem.* 1991,*196,* 459-467.
23. Schinbinger, M. F.; Redl, B., Stoffler, G. *Biochem. Biophys. Acta.* 1991, *1073,* 132-148.
24. Zwickl, P.; Fabry, S.; Bogedain, G.; Haas, A.; Hensel, R. *J. Bacteriol* . 1990, *172,* 4329-4338.
25 Courchesne W.E.; Magasanik, B. *J. Bacteriol.* 1988, *170,* 708-713.
26. Miller, S.M.; Magasanik, B. *J. Bacteriol.*1990.*172,* 4927-35.

RECEIVED January 15, 1992

Chapter 7

Biocatalysis in Organic Media

Don A. Cowan[1] and Adrian R. Plant[2]

[1]Department of Biochemistry and Molecular Biology, University College
London, Gower Street, London WC1E 6BT, United Kingdom
[2]Sigma Chemical Company, Fancy Road, Poole, Dorset BH17 7NH,
United Kingdom

It is now accepted that organic media are widely applicable to
biocatalytic processes. Enzyme reactions have been demonstrated in
many different solvent systems containing apolar solvents with a wide
range of hydrophobicities. In such systems the aqueous fraction of the
solvent may vary from trace levels (the "microaqueous" reaction
system) to the major constituent.

The properties of enzymes in media containing both aqueous
and apolar constituents are influenced by many factors including the
dielectric constant, the concentration or "activity" of water, the presence
of phase-interfaces, etc. All of these factors influence the structural
stability of the enzyme to some degree and optimization of enzyme-
stability in response to these factors is of critical importance in the
design of any organic-solvent biocatalytic system.

The presence of apolar solvents, even at low concentrations,
may influence protein structure at a more subtle level, resulting in
significant changes in reaction kinetics, substrate specificity and
stereoselectivity. Thus "solvent engineering" can be used to exert
control over reaction mechanisms, product spectrum and yield

In order to take full advantage of the effects of different
solvents on enzyme function in designing reaction systems, an
understanding of the molecular interactions between solvent and protein
is essential. For example, a detailed knowledge of the role of protein
solvation shell water and the distortion of this shell by hydrophilic
organic solvents may assist in the design of "solvent resistant"
enzymes. Similarly, the inferred influence of low water content
solvents on conformational mobility may permit us to select solvent
systems which will optimise the chemical and optical purity of the
reaction products.

Organic Phase Reaction Systems

Introduction. Numerous factors are important in defining the physical and chemical
effects of any aqueous-organic solvent system on constituent proteins. Nevertheless,
in order to optimise or indeed establish a viable organic phase biocatalytic system, it is
important to be aware of the manner in which these factors impinge on biocatalyst
structure and function. Single properties such as organic phase hydrophobicity and

0097–6156/92/0498–0086$06.50/0

dielectric constant, and system characteristics such as interfacial area and interfacial tension, can be more or less critical determinants of enzyme viability. The way in which these factors are determined by the type of solvent system will be stressed in this chapter.

While it is popular to define solvent systems (such as miscible, immiscible, microaqueous, etc) as unique systems, these are merely broad approximations of the more obvious physical characteristics of the system. Moreover, to a large extent these "solvent systems" are defined points on a composition continuum determined by two primary components: organic phase hydrophobicity and water content.

Organic Phase Hydrophobicity. One of the primary questions for any researcher intending to use an organic phase reaction system is "which organic solvent should I choose?". The criteria for selection will include numerous variables such as substrate and product partitioning characteristics but one of the most critical factors is the optimization of biocatalyst stability. Thus over the past ten years much effort has been expended on a search for methods of ranking organic solvents in relation to their effect on protein stability in organic:aqueous solvent systems. The incentive is to provide a viable indicator of solvent applicability, a valuable determinant for both research and industrial operations.

Organic phase hydrophobicity has proved to be the most useful factor. However, there are a number of different ways of expressing this property (Table I), not all of which prove to bear any useful relationship with the effect of the solvent on protein stability. For example, Laane *et al.* [1] have shown that there is little obvious correlation between the values of δ, E_T, ε and μ and the retention of biocatalyst activity for a range of organic solvents in organic-aqueous systems. Conversely, log P has proved to be a particularly valuable determinant of the feasibility of using particular organic phase components in organic phase biocatalysis [2].

Table I. Chemical Indices of Hydrophobicity

Symbol	Parameter
δ	Hildebrand solubility parameter
E_T	Dye solvatochromism
ε	Dielectric constant
μ	Dipole moment
Log $S_{w/o}$	Saturated molar solubility
log P	Organic-aqueous partition coefficient

Source: From refs. [3] and [4].

Log P is defined as the logarithm of the partition coefficient of a substance between octanol and water (equation (1)). It is determined either experimentally or (for simple molecules) from *hydrophobic fragmental constants* [5]. The values of log P for a number of common solvents are given in Table II.

$$P = \frac{[\text{solute}]_{octanol}}{[\text{solute}]_{water}} \tag{1}$$

Studies of the effects of organic solvents on stability of intracellular enzymes has led to the much quoted generalisation that solvents of log P > 4 are generally suitable for use in organic phase systems, those of log P = 2 - 4 are of specific use

Table II. Log P values for common organic solvents

Solvent	Log P (at 25 deg.C)	Solvent	Log P (at 25 deg.C)
Dimethylsulfoxide	- 1.35	1-chloropropane	1.82
Methanol	- 0.79	Heptan-2-one	1.82
Dioxane	- 0.47	Hexanol	1.86
N,N-dimethyl formamide	- 0.42	Dipropyl ether	1.94
Acetonitrile	- 0.36	Benzene	2.00
Acetone	- 0.30	Chloroform	2.20
Ethanol	- 0.26	Methoxybenzene	2.15
Ethoxyethanol	- 0.22	Methyl benzoate	2.16
Methylacetate	0.11	Pentyl acetate	2.13
Propan-2-ol	0.15	Octan-2-one	2.35
Butan-2-one	0.24	Heptanol	2.39
Propan-1-ol	0.27	Toluene	2.60
Tetrahydrofuran	0.49	Ethoxy benzene	2.68
Ethylacetate	0.65	Ethylbenzoate	2.69
Methylpropionate	0.64	Dibutylamine	2.76
Butan-2-ol	0.68	Chlorobenzene	2.81
Pyridine	0.71	Octan-2-ol	2.90
Pentan-3-one	0.77	Carbon tetrachloride	2.98
Pentan-2-one	0.76	Pentane	2.99
Butan-1-ol	0.80	Styrene	3.00
Diethyl ether	0.92	Xylene	3.10
Cyclohexanone	0.94	Ethylbenzene	3.12
Propylacetate	1.17	Cyclohexane	3.20
Pentan-2-ol	1.21	Nonanol	3.45
Pentan-3-ol	1.21	Hexane	3.52
Ethyl chloride	1.29	Methyl cyclohexane	3.59
Hexanone	1.29	Propylbenzene	3.65
Pentan-1-ol	1.33	Decanol	3.98
Phenol	1.50	Heptane	4.05
Triethylamine	1.54	Diphenylether	4.06
1,2-dichloroethane	1.63	Undecanol	4.51
2-chloropropane	1.70	Octane	4.58
Butylacetate	1.70	Dodecanol	5.04
Nitrobenzene	1.81	Dibutylphthalate	5.40
Heptan-3-one	1.82		

Source: Data taken from references [4] and [111].

(depending on the stability of the enzyme for example) and those of log P < 2 are of little or no value. However, it must be appreciated that these empirical rules have been established using whole cells rather than purified enzyme preparations and are derived from study of a limited range of biocatalysts. Recent work has indicated a number of examples where this correlation is a less than accurate guide to the selection of the organic phase component.

The use of log P as a definitive indicator of relative hydrophobicity in high temperature organic phase biocatalytic systems may be even less reliable. Log P is a temperature-dependent parameter where changes in both the extent and direction of values will be related to the relative thermodynamics of dissolution in octanol and water.

The value of log P as an indicator of solvent suitability has been further supported by the establishment of a good correlation [6] between this index and the purely empirical "threshold solvent concentration" (C_{50}; the organic solvent concentration at which 50% inactivation of the enzyme is observed). These values were derived from the denaturation profiles of purified enzymes in miscible organic:aqueous solvent systems and, while being limited to single phase solvents, do at least demonstrate that log P might be equally applicable to purified enzymes as to cell-encapsulated enzymes.

Nomenclature of Organic Phase Systems. In the previous section, we have referred to various solvent systems including "miscible", "immiscible", "single phase", "biphasic" etc. and some explanation of the nomenclature, structure and relationships between these different solvent "categories" is worth including. Organic phase reaction systems are generally described on the basis of broad physical characteristics such as solvent miscibility. This has led to the common designation of "single phase", "biphasic" and "multiphasic" systems (as represented in Figure 1). In reality, however, the difference between these systems (particularly the single and biphasic) is more subtle and changes in water content and/or organic phase hydrophobicity yield a continuum between the two extremes. This continuum can be represented in the form of a phase diagram (Figure 2). The physical and chemical characteristics of these various solvent systems will be discussed in more detail below.

Two alternative non-aqueous monophasic reaction systems which cannot strictly be classified as "organic solvent" are supercritical fluid and gas-phase reaction systems. These will be discussed in more detail below.

"Anhydrous" or Microaqueous Solvent Systems. The term "anhydrous" is something of a misnomer in that these solvent systems are characterised by the presence of **very low** rather than **zero** water contents. Although the water content of organic solvent component can be reduced to low levels by fractional distillation or by adsorption with anhydrous calcium sulphate or molecular sieves, it is practically impossible to remove the "bound" water from the protein component. Thus the term "microaqueous", proposed by Yamane [7,8], is a more accurate representation of the real state.

In general, the "microaqueous" solvent systems used experimentally are rather poorly characterised in terms of the water content. Assessment of the true water concentration of the microaqueous state is not helped by the experimental difficulty of assessing the water contents of organic solvent systems [9]. The normal experimental approach is to desiccate the organic solvent, to assess the protein water content by Karl-Fischer analysis or some similar technique, and to subsequently specify the water content of the system by addition of an accurately defined volume of aqueous solvent. This approach provides no enlightenment on the molecular distribution of water in such systems. Yamane [8] has derived a relationship from which the protein-bound water (y), on addition of a small volume of aqueous solvent to a protein-containing

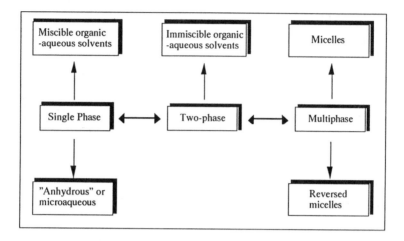

Figure 1. Schematic representation of organic-phase solvent systems.

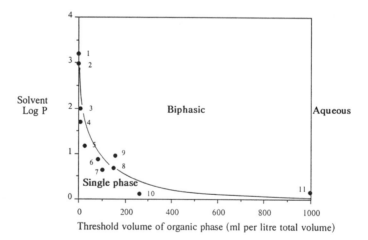

Figure 2. "Phase diagram" representation of organic-aqueous solvent systems showing the threshold volumes of organic phase required for formation of a biphasic system. [Data points are: 1, cyclohexane; 2, pentane; 3, benzene; 4, butyl acetate; 5, propyl acetate; 6, diethyl ether; 7, ethyl acetate; 8, butan-2-ol; 9, cyclohexanone; 10, methyl acetate; 11, propan-2-ol.]

organic phase, may be determined (equation (2)). However, the validity of this value is still dependent on several experimental quantities of uncertain accuracy, namely h_p and h_s, the water contents of solvent and protein powder and $C_{total\ water}$ and $C_{free\ water}$, the concentrations of total and free water. W_p and W_s are the weights of dry protein powder and organic solvent respectively.

$$y = \frac{C_{total\ water}\,(1\text{-}h_p)W_p + 100\{(C_{total\ water}\text{-}C_{free\ water})/(100\text{-}C_{free\ water})\}.(1\text{-}h_s)W_s}{(100\text{-}C_{total\ water})(1\text{-}h_p)W_p} \times 100 \quad (2)$$

Proteins are almost wholly insoluble in "microaqueous" organic solvents. The only exceptions to this are the extremely hydrophilic organic solvents such as dimethylformamide and pyridine. However, solubility does not necessarily imply loss of native structure and hence function. Indeed, early work by Dastoli and Price [10], subsequently extended by Klibanov [11] has demonstrated that enzymes are capable of retaining their native conformation in such systems with surprising (and sometimes favourable) consequences in terms of activity, specificity and stability. The molecular basis of these properties is discussed in more detail below.

Proteins retain solubility in microaqueous and low water content solvent systems only where the solvent is sufficiently solvating (as in the case of dimethylformamide) or where the protein has been chemically modified to enhance the surface hydrophobicity. In the former case, an inspection of log P values of pyridine and dimethylformamide would suggest that proteins, while solubilized by these solvents, should be inactivated very rapidly. While this is probably true in most instances, the achievement of functional enzyme reaction systems in DMF and similar solvents has been reported [12]. Notably, the enzymes used in these reactions have often been intrinsically stable or have been further stabilised by site-directed mutagenesis.

Organic solvent solubilisation of proteins by chemical modification has been successfully achieved by covalent attachment of polyethylene glycol via free amino groups, often to high substitution levels. This has resulted in a protein product which is both soluble and active in hydrophobic organic solvents such as benzene [13,14] and dimethylformamide [15].

Miscible Organic-aqueous Solvents. A protein dissolved in a single phase organic-aqueous solvent is directly exposed to the organic component. The consequences of this exposure (in terms of protein stability) depend on the organic solvent concentration and the chemical interaction between (i) the solvent and the protein surface groups and (ii) the solvent and the protein hydration shell. Since in order for an organic solvent to be fully miscible with the aqueous phase the solvent log P must be fairly low, such systems are often detrimental to protein stability at all but low organic solvent concentrations. For example, α-chymotrypsin exhibited 50% inactivation in dimethylformamide, ethanol and isopropanol, at 26%, 33% and 23% vol:vol ratios respectively [6]. A general solvent response was also noted where the enzyme retained full stability until a critical organic cosolvent concentration was reached. With a further increase of only 2-3% in cosolvent concentration, enzyme stability was reduced abruptly [6]. This emphasises the need for careful design of single phase organic-aqueous solvent systems.

Immiscible Organic-aqueous Solvents. Stirred immiscible organic-aqueous solvent systems are quite commonly used in industrial processing and are characterised by one feature not found in other solvent systems: the presence of liquid:liquid interfaces. In other respects, the biphasic solvent system is not dissimilar to that found in a single phase organic-aqueous solvent. Depending on the aqueous solubility of the organic solvent, the biocatalyst residing in the aqueous phase will "experience" a greater or lesser concentration of organic solvent (see Table III for aqueous solubility data for various organic solvents). The hydrophilicity of the organic solvent thus influences it's aqueous solvent concentration and, presumably in some parallel manner, it's degree of interaction with protein-bound water which is a critical determinant of protein stability. Where higher temperatures significantly reduce aqueous solubilities (e.g., cyclohexanone, where the solubility in water at 20°C is 1.53 mol.l^{-1} cf 0.51 mol.l^{-1} at 50°C), increases in protein instability may be significantly less than might be predicted from temperature/solvent effects alone.

The presence of liquid-liquid interfaces is generally detrimental to protein stability. Interfacial denaturation of proteins is well known, particularly at gas-liquid interfaces, although the molecular mechanisms are not particularly well understood. In liquid:liquid systems, little information is available on the importance of interfacial area and the factors which influence interfacial denaturation. Various studies by Lilly and others [16] indicate that in stirred biphasic solvent systems, stirring rate (which is related in a complex fashion to interfacial area) is positively correlated with enzyme instability. A general decrease in protein stability was also observed [17] as the organic:aqueous phase ratio was increased when the organic phase was n-hexane but the effect was much less significant when the organic phase was n-butanol. This gives some indication of the relative importance of interfacial denaturation versus "intra-aqueous" denaturation demonstrating that, where the organic solvent has a very low aqueous solubility, interfacial effects can be observable and significant.

There are also indications that interfacial tension may play a significant role in protein denaturation. Interfacial Tension (γ_i: Nm^{-1}) is a fundamental property of interfaces. At a molecular level this *tangential stress* is a result of a contractile tendency at the interface between two immiscible liquids, resulting from orientated packing of molecules at the interface [18]. Interfacial tension is directly related to interfacial energy where molecules at interfaces have greater energy than those in the bulk phase [19]. Interfacial tension is greatly influenced by the presence of other compounds (buffers, proteins, substrates and products, other solvents) and by temperature (equation (3)):

$$\gamma = \gamma_0(1\text{-}T/T_c)^{1.23} \qquad\qquad (3)$$

This suggests that higher temperatures will generally reduce interfacial tension although the extent of this effect may not be great.

The involvement of interfacial tension in enzyme stability is implicated in a comparison of the kinetics of protein denaturation in a limited range of solvents with similar log P values [17]. In each instance, the denaturant effect was greater with the solvent having the higher nominal interfacial tension value (Table IV). However, those solvents showing enhanced denaturation are also chlorinated hydrocarbons, and this correlation should be viewed with some care. Nevertheless, these indications of a potentially useful selection criterion justify further investigations of the influence of interfacial tension.

Reversed Micelles. An immiscible aqueous phase can be supported in a bulk organic phase by the addition of amphipathic molecule such as AOT (sodium 1,2-bis(2-ethylhexyloxycarbonyl)-1-ethane sulfonate) or CTAB (cetyl trimethylammonium bromide). These "water-in-oil" microemulsions are simple to generate and particularly convenient for biochemical studies, being thermodynamically stable over a wide range of temperatures and constituents, and possessing optical clarity. The micellular size can be readily varied between 50Å and 300Å depending on the composition. The preparation and application of reversed micelles has been reviewed in considerable detail [20,21] and there are numerous recent publications on the stability and activity of encapsulated enzymes (see reference [21] for a comprehensive list of references).

Table III. Molar solubilities of some common organic solvents in water

Solvent	Solubility in water at 20°C	Solvent	Solubility in water at 20°C
Benzene	0.073	Heptan-2-one	0.038
Butan-1-ol	0.98	Heptane	0.00017
Butan-2-ol	1.64	Heptanol	0.015
Butan-2-one	3.14	Hexane	0.00016
Butyl acetate	0.058	Hexanol	0.069
Carbon tetrachloride	0.005	Methyl acetate	3.25
Chlorobenzene	0.0043	Octan-2-ol	0.0098
Chloroform	0.068	Octan-2-one	0.0088
Cyclohexane	0.0012	Octane	7.4×10^{-6}
Cyclohexanone	1.53	Pentan-1-ol	0.31
Decanol	0.00025	Pentane	0.0005
Dibutylamine	0.036	Propan-1-ol	13.31
Dibutylphthalate	0.00036	Propan-2-ol	13.00
Diethyl ether	0.795	Propyl acetate	0.225
Dipropyl ether	0.048	Pyridine	12.37
Ethyl chloride	0.089	Toluene	0.0056
Ethyl acetate	1.02	Triethylamine	0.532
Ethylbenzene	0.0017		

Source: Data taken from The Merck Index, 10th Edition (Merck & Co. Inc.) 1983 and reference [109].

Table IV. Interfacial tension as a possible factor in protein instability

Solvent	Log P	Interfacial Tension $(Nm^{-1} \times 1000)$	Protein Stability[a]
n-Hexanol	2.0	6.8	72 ± 27 [5]
Chloroform	2.0	23.0	15 ± 10 [5]
n-Octanol	2.9	8.5	79 ± 29 [5]
Tetrachloromethane	3.0	45.1	38 ± 23 [5]

[a] Protein stability is specified as the percentage total protein not precipitated after incubation in a stirred buffer/organic solvent mixture at 50°C for 1 hour.

Despite extensive evidence for enhanced protein stability [e.g., 22,23] and superactivation of enzymes [24,25], little is known of the physical state of enzymes in reversed micelles. The currently accepted model assumes a spherical boundary layer of detergent molecules encapsulating an aqueous phase in which the protein is protected from the bulk organic phase [20]. The mediation of the detergent monolayer between the interior protein and the exterior organic phase is undoubtedly an important factor.

Stable microemulsions which lack the detergent component can be generated by using a ternary solvent systems such as hydrocarbon/isopropanol/water [26]. Like reversed micelles, these are thermodynamically stable and optically clear. They have the potential advantage that recovery of reaction products is not necessarily hindered by the presence of the surfactant molecules.

Supercritical Fluid Systems. Supercritical fluid biocatalysis [recently reviewed in reference 27] is an extension of the "microaqueous" solvent reaction system. A supercritical fluid is any compound above its critical temperature and with pressure above the corresponding critical value. The properties of supercritical CO_2 (temperature above 32°C and pressure above 72 bar) are particularly appropriate for biological processes and supercritical CO_2 has already found major industrial application as an extractant in the production of essential oils, fragrance substances, hop extracts and in decaffeination. Particular advantages of supercritical gasses are their inert, non-toxic properties, high diffusivity, low viscosity and low surface tension. Solute solubility is very dependent on temperature and pressure, providing a convenient mechanism for the downstream separation of unreacted substrates, products and enzymes. The low solubility of water in supercritical CO_2 also provides a subtle control of water content (and water activity), by which reaction equilibria may be controlled. A number of biocatalytic reactions, including esterification, hydrolysis and oxidation [27] have been studied in supercritical fluids. From the limited data available, it seems that enzymes are no less (or more) unstable in supercritical fluids than in organic systems [e.g., 28]. It has been reported that reaction rates are faster in supercritical fluids than in organic or microaqueous liquids [27,29].

Gas-phase Bioreactors. The realisation that enzymes were capable of retaining catalytic function in the absence of a bulk water phase has inevitably led to consideration of other non-aqueous reaction systems such as the gas-solid reactor. Although gas-liquid reaction systems (where one of the substrates is a gas such as O_2 or CO_2) have been investigated [30-32], there have been few experimental investigations of the feasibility of reactions involving wholly gaseous reactants and "anhydrous" enzymes. Hou [33] was able to demonstrate propylene oxide production from Methylosinus cells immobilised onto porous glass beads held in a stream of propylene. The reaction utilised an NAD-dependent methane monooxygenase and the cofactor was regenerated by periodically replacing substrate with a flow of methanol vapour. Propylene oxide production decreased rapidly over several reaction/regeneration cycles due probably to the lability of the cofactor. Similarly, it has been shown [34] that the alcohol oxidase of immobilised yeast cells could effect gas-phase catalysis of acetaldehyde to ethanol. Pulvin et al.[35] coimmobilised NAD(H) and alcohol dehydrogenase into albumin-glutaraldehyde porous particles. The steady-state reduction of acetaldehyde and cofactor regeneration was limited by the operational stabilities of the enzyme and cofactor, which were similar to those found for the reaction in aqueous systems.

The gas-phase conversion of oxygen and ethanol to acetaldehyde and hydrogen peroxide has been studied in limited detail [36,37] using alcohol oxidase from Picia pastoris immobilised onto DEAE cellulose. Gas phase biocatalysis was strongly dependent on the water activity of the system and on product inhibition by H_2O_2. The latter was readily removed by coimmobilisation of catalase or peroxidase. The dried enzyme preparation was significantly more thermostable than the aqueous enzyme.

Previous experiments on gas-phase biocatalysis have used an extremely simple enclosed reaction vessel at temperatures of up to 50°C. Within a relatively short time, it is probable that more sophisticated gas-flow systems will appear, utilising higher reaction temperatures and a much wider range of enzymes.

The Effect of Organic Solvents on Protein Structure

Changes in Conformation and Protein Denaturation. Much has been made of the fact that certain enzymes have shown quite startling increases in thermostability when suspended in "microaqueous" organic solvents [38-40]. However, this is a specialist case and the manner in which the organic solvent component of either biphasic or monophasic organic-aqueous solvent systems effect the structure and function of proteins is of more general interest.

Non-aqueous solvents undoubtedly induce changes in protein conformation. Studies with poly-amino acids (discussed in detail by Singer [41]) indicated that many of the more hydrophilic solvents increased α-helical content. This has also been observed with globular proteins in such solvents as methanol [42], 2-chloroethanol [43], and dioxane [44]. A model [equation (4)] for solvent-induced conformational changes in globular proteins was developed by Herskovits [42] but is probably applicable only to solvents with a low log P.

$$N \Leftrightarrow I \Leftrightarrow H \qquad\qquad (4)$$
$$K_a \quad\ K_b$$

N, I and H are the native, intermediate and helical forms respectively and K_a and K_b are the equilibrium constants for the first and second transitions. In this model, the solvent first converts the protein to the intermediate form with a loss of some degree of globular state, a process evidenced by changes in absorbance characteristics. It is probable that loss of biological function also occurs at this point. The second transition results in the total loss of globular structure and the appearance of an unfolded form rich in α-helical content. This is significantly different from the mechanism of protein denaturation by chaotropic agents such as guanidinium hydrochloride and urea where the process involves a two-state transition leading to a random coil [45].

The molecular mechanisms of protein denaturation by organic solvents remain unclear. In thermodynamic terms, the conversion of a stable folded form to a stable unfolded form can involve destabilisation of the folded conformation and/or stabilisation of the unfolded form. The effect of various short chain aliphatic and aromatic alcohols on the thermal transition of ribonuclease was interpreted in terms of stabilisation of the unfolded form via hydrophobic interactions between the solvent molecules and hydrophobic amino acid residues [46]. However, a tendency for the more hydrophilic organic solvents to induce protein denaturation more readily than those of lower hydrophilicity argues against this mechanism as the most important determinant of destabilisation. A more critical factor may be the tendency for hydrophilic (low log P) solvents to interfere with the hydration shells of surface residues. This is described by Khmelnitsky [47] in terms of their capacity to maintain "solvophobic" interactions, that is, to replace water in the hydration shell of a protein without significant distortion of hydrophobic interaction within the globular structure.

Just how many water molecules are required to maintain a globular protein in its native conformation is not known. It has been calculated that a molecule of chymotrypsin in "anhydrous" octane retains fewer than 50 molecules of water [11], considerably less than the number required to provide a molecular monolayer over the protein surface. For other enzymes [38], several hundred water molecules per enzyme molecule appear to be necessary for the onset of activity. This difference may reflect the nature (size or mobility) of the active site and/or the characteristics of the catalytic mechanism.

Protein Conformational Mobility. As pointed out by Klibanov [11], water is critical to the normal conformational mobility of proteins and to the extended conformational processes leading to protein denaturation. The current view is that removal of the bulk water phase restricts global conformation mobility (i.e., enhances protein rigidity) with the consequence that the energetic barriers to protein unfolding are increased and the protein exhibits higher thermostability.

Support for this view comes from a body of data [17] indicating that the organic solvent stability and thermostability of proteins are closely correlated (Figure 3). A variety of physical studies [e.g., 48] have demonstrated that proteins of higher thermostability have reduced conformational mobility, thus implying that the same may apply to proteins exhibiting stability in organic solvents.

A number of enzymes are activated, sometimes by more than 10-fold, by the addition of miscible organic solvents at concentrations of 10-30% [49-51]. One possible mechanism for this activation is a relaxation of conformational flexibility [47]. A corresponding decrease in the structural stability of the enzymes should be expected under these conditions.

The Effect of Protein Immobilisation. There are a number of reports that enzymes covalently immobilised to insoluble matrices are significantly stabilised with respect to inactivation and denaturation in organic solvents (see references in [47]). The data is currently insufficient to indicate whether this might be a general trend, but examples of immobilisation reducing protein stability have been reported [52-54]. Mechanistically, enhanced stabilisation could be brought about by a reduction in conformational flexibility. The presence of multi-point covalent linkages between protein and matrix presents an obvious means for the restriction of complete protein unfolding. However, other mechanisms such as an increase in the concentration of free water in the region of the immobilized enzyme could generate a similar effect.

Support for multi-point stabilisation is seen in a study (Sergeva, M.V., 1988; cited in reference [47]) where the stability of immobilised α-chymotrypsin in dimethylformamide, aliphatic alcohols and diols was increased as the number of covalent linkages between enzyme and matrix was increased.

The Effect of Solvent on Enzyme Function

Equilibrium Effects. Studies of the effects of organic solvents in biocatalytic media have focussed primarily on the biotransformation of compounds with low solubility in water or the promotion of reactions which are thermodynamically unfavourable in aqueous systems. For example, in aqueous media the equilibria of hydrolytic reactions catalysed by hydrolases lies strongly in favour of the hydrolytic fragments. However, in organic solvents the equilibria may be shifted to give a useful yield of synthetic product because of it's preferential solvation of the organic phase relative to the fragments produced by hydrolysis. The required equilibrium shift may be achieved by adding water-miscible solvents to aqueous media, performing reactions in "neat" solvents or by employing a biphasic system if no one solvent can provide a significant equilibrium shift [55].

The biosynthetic applications of solvent induced equilibrium shifts are exemplified by the use of proteases for peptide synthesis from protected amino acids [56-58] or by the esterification and transesterification reactions of esterases and lipases

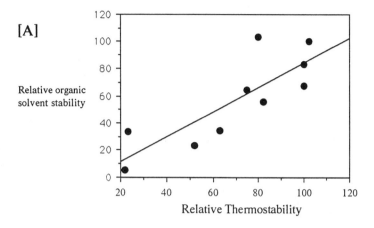

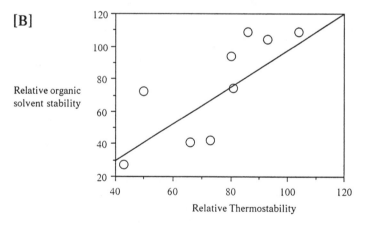

Figure 3. Correlation between protein thermostability and organic-solvent stability for [A] protein populations and [B] purified proteolytic enzymes.
[For [A], the organic solvent stability and thermostability values were estimated as percent protein not precipitated after incubation for a specified period in 50:50 buffer: n-butanol and aqueous buffer, respectively. For [B], organic solvent stability and thermostability values were estimated as the residual enzymic activity after incubation under conditions as described above.]

which have been utilised for the synthesis of many hydrophobic esters in high yield [59-62].

Although thermodynamic considerations can enable the equilibrium shift to be predicted, they do not of course predict the rate at which equilibrium is attained. For most hydrolases, the hydrolytic rate is generally faster than the corresponding rate for synthesis in organic media. Yet the kinetic restraints of the latter reaction could in principle be overcome by performing the synthetic reaction at very high temperatures in low-water content media. The activity of "dry" porcine pancreatic lipase at 100°C exceeds that at 20°C by a factor of 5 [39]. The hydrolytic reaction can not be performed at such high temperatures which lead to rapid denaturation of the enzyme.

It is important to recognise that water produced as a product of reversed hydrolysis reactions in microaqueous media may adversely effect the reaction equilibrium unless it is removed, e.g., by inclusion of molecular sieves or by bubbling dry air through the system [63,64]. Ergan et al. [65] esterified glycerol and oleic acid with Mucor miehei lipase at elevated temperatures (up to 80°C). Increasing the temperature shifted the equilibrium in favour of ester synthesis by facilitating evaporation of water and also shortened the time needed to achieve maximum yield. These experiments involved enzymes suspended solely in a reaction medium composed of substrates, without added solvents, and indeed it is sometimes possible to help shift reaction equilibria by chosing one or more substrates which also serve as solvents. Thus ethyl acetate has been used as an acyl transfer agent and solvent to acylate β-substituted ethanols using Candida cylindracea lipase [56] and acrylate monoesters of various diols have been synthesized using Chromobacterium viscosum lipase suspended in ethyl acrylate [66].

The synthetic reaction catalysed by lipases may also be promoted by using irreversible acyl transfer agents such as oxime esters [67] or more frequently, enol esters [60]. This results in the effective removal of the enol product which undergoes tautomerism to an aldehyde or ketone and is unavailable to the back reaction. The use of elevated reaction temperatures also offers the possibility of removing relatively volatile ketones and aldehydes from the reaction. Organic reaction media may facilitate the use of unusual water-labile substrates such as acid anhydrides which are suitable irreversible acyl transfer agents for many lipases but which are rapidly destroyed by water. Acid anhydrides have been used for lipase mediated acylation of nucleosides in DMSO, DMF, or DMA [68], and a range of alcohols in diethyl ether [69] or benzene containing a very low water content [70].

It is possible to influence reaction equilibria favourably by using water miscible solvents to perturb the ionic equilibria of the reactants. The thermodynamic barrier to peptide synthesis from free amino acids would be lowered considerably in both reactants and products existed in non-ionized forms [71]. Unfortunately, the amphoteric natures of amino acids and peptides ensures that they are at least partially ionized at all pH values. Inclusion of a water miscible solvent lowers the dielectric constant and tends to dehydrate ionic groups resulting in reduced acidity of carboxyl groups and lower basicity of amino groups thus abolishing partially the ionic properties of amino acids and peptides.

Elevated reaction temperatures may also be used to perturb ionic equilibria and so favour peptide synthesis. While the ionization of carboxyl groups is little influenced by temperature, the pK_a values of amino acid α-amino groups decrease markedly with increases in temperature [71,72] and it is likely that biocatalytic synthesis of peptides would be enhanced by using stable proteases at very high temperatures in organic phase reactors.

Chemical Selectivity and Regioselectivity. An important consequence of the use of enzymes at low water content in organic solvent biocatalysis is that changes in chemical specificity, regioselectivity and the ability to discriminate between optical antipodes of the substrate are often seen. Enzymes with very narrow substrate specificity in aqueous systems probably have restricted conformational mobility and are relatively rigid structures, at least in the region of the active site. Such enzymes have high energetic barriers to the disruption of the non-covalent forces influencing their conformational states and are likely to be only minimally influenced by the transition from aqueous to apolar media [73]. Accordingly, it would be expected that differences in substrate specificity and enantioselectivity would be slight in aqueous and apolar solvents. In contrast, the conformation of enzymes with a broad substrate specificity is presumably less constrained and the binding of substrate is effected in a manner similar to that of "induced fit" enzymes [74,75]. This type of enzyme would have lower energetic barriers to perturbation of its conformation which would be expected to be different in aqueous and more hydrophobic media. Thus "induced fit" enzymes should exhibit changes in chemical specificity, regioselectivity and enantioselectivity associated with conformational transitions experienced in different media. Similarly, the allosteric or cooperative properties of some proteins are likely to be effected by solvents as these phenomena are intimately linked with conformational mobility [76].

Where the phenomenon has been investigated, it has often been found that substrate specificity is narrowed in organic media but enantioselectivity is increased [77,78]. This rather generalized conclusion can be explained if the substrate binding site of a conformationally flexible aqueous enzyme can be readily induced to adapt to and "fit" a wider variety of substrates than a more conformationally rigid enzyme in apolar media. In many organic solvents, the substrate adopts a more precise spatial orientation with respect to the catalytic centre of the more rigid binding site and increased enantioselectivity is a consequence of a spatially restrained microenvironment imparting a greater directional component to the catalytic sequence.

Porcine pancreatic lipase (PPL) is able to transesterify tributyrin with a range of alcohols in media containing 0.7% water. If the water content is reduced to 0.015%, the enzyme is unable to utilise tertiary alcohols as it lacks the conformational flexibility needed to accommodate the bulky substrates at it's active centre [39]. Interestingly, the dry lipase remained inactive to tertiary alcohols even at 100°C but in the presence of 0.7% water it was active on the same substrates at 20°C. Clearly the influence of hydration on conformational mobility and catalytic activity was greater than the influence of kinetic energy due to heating between 20°C and 100°C. Steric factors are probably important in the aqueous hydrolysis of alkyl butyrates by PPL and Pseudomonas fluorescens lipases. The hydrolytic reaction favoured large C_{18} alcohol substituents but this preference was lost when the corresponding transesterification was performed in diisopropyl ether or hexane while in acetonitrile, the reaction became more specific for shorter carbon chain length reactants [78].

Organic solvents can also effect substrate specificity by altering the affinity of the substrate binding site for its ligand resulting in changes in the apparent K_m and the catalytic efficiency or "specificity" constant k_{cat}/K_m [73,80]. In aqueous solution the polar groups of a substrate binding site form hydrogen bonds with water. Binding of the substrate disrupts some of these hydrogen bonds which are replaced by interactions between the substrate and polar groups of the binding sites [81]. In the presence of a hydrophobic non-hydrogen bonding solvent the replacement of active site water by substrate is a major obstacle to substrate binding [82] resulting in a decrease in affinity of an enzyme for its substrate. Active site bound water is in equilibrium with water in the bulk solvent phase and is more readily dissociated from the enzyme if the bulk phase is a polar hydrogen-bonding solvent. In general, changes in substrate binding affinity occurring on transition from a polar to a non-polar environment will be governed by the relative partitioning of the substrate and water between the active site and bulk solvent phase. Thus α-chymotrypsin which exhibits low K_m values for hydrophobic amino acids in aqueous solution will preferentially bind hydrophilic amino acids in apolar media. In octane k_{cat}/K_m for the hydrolysis of N-acetyl-L-phenylalanine ethyl ester is 20 times less than for the histidine derivative while in aqueous solution k_{cat}/K_m for the phenylalanine derivative 100 times higher than for the histidine derivative [83].

Pronounced substrate or product inhibition is sometimes apparent on transition from an aqueous to a non-hydrogen bonding environment if the substrate or product is highly polar. Under such conditions the apparent association constant of enzymes with substrate or product is so large that the ligand is effectively irreversibly bound [73].

Although changes in substrate specificity consequent of altered polarity of the medium had been reported for many enzymes including PEG-modified thermolysin [84], α-amylase [85], α-thrombin [86], lipase [79] and peroxidase [80], it is not always easy to distinguish the effects of perturbations of hydrogen bonding and hydrophobic interactions from steric effects arising through conformational changes. Indeed solvent-induced conformational changes will vary the number and nature of possible hydrogen bonding and hydrophobic interactive sites available to the enzyme and its ligand.

Preparations of subtilisin [82] and chymotrypsin [38] lyophilised from aqueous solution in the presence of ligand were more active catalytically in organic solvents and had differing specificities from preparations lyophilised in the absence of ligand. Dehydration in the presence of ligand effectively "locked" the enzymes in an active conformation. Furthermore, the presence of ligand at the active centre was reasoned to have prevented hydration of polar substrate-coordinating groups and thus to have obviated the need for the substrate to disrupt hydrogen bonds on binding with the enzyme.

Changes in regioselectivity of the enzymic catalyst in the presence of organic solvents have been reported for a variety of enzymes including nitrilase [87], protease [89] and lipase [88,89]. Selective deacylation of 1,6-anhydro-2,3,4-tri-O-butyryl-β-D-glucopyranose by PPL, wheat germ lipase and chymotrypsin was strongly influenced by the concentration of methanol in an aqueous system [89]. On going from 20% to 50% methanol preferential deacylation moved from C4 to C3 for chymotrypsin, from C2 to C4 for PPL and from C2 to C3 for the wheat germ enzyme. However, the selectivity of the deacylation of 1,6-anhydro-2,3,4-tri-O-butyryl-β-D-galactopyranose catalysed by PPL and Candida cylindracea lipases was almost completely lost in the presence of 1.25% n-butanol [90]. Changes in regioselectivity in the presence of solvents probably reflect subtle changes in hydrogen-bonding/conformation similar to those responsible for alteration of gross chemical specificity as discussed above.

The greater temperature range available to solvent biocatalysis by virtue of enhanced enzyme stability in hydrophobic media is a factor for consideration by the "solvent engineer" seeking to promote useful modification of biosynthetic reactions. At ambient temperatures the condensation of diols with diacids catalysed by lipases in isooctane resulted in a preponderance of linear oligomeric esters [91,92] but at 50°C to 75°C dilactones were the major product. Presumably elevated temperatures enabled the barrier imposed by the higher energy of activation for the lactonization to be overcome.

Stereoselectivity Effects. Enzymic catalysis in aqueous media is often selective for one enantiomer of a racemate, a particular enantiotopic or diastereotopic group or a specific face of a prochiral molecule [93]. These properties have been central to the development of biotransformations as routes to useful chiral synthons. Selectivity is often maintained, enhanced or reversed in organic solvents. Using these characteristics, chloroperoxidase has been used to asymmetrically oxidise sulphides [94], racemic hydroperoxides have been resolved with lipoprotein lipase [95], phospholipids containing a chiral phosphorous have been synthesized with phospholipase A_2 [96], prochiral ketones reduced to chiral alcohols with alcohol dehydrogenases [97] and aldol condensations using aldolase [98] have been performed in the presence of organic solvents.

Wong [93] argued that stereoselectivity often originates from the energy difference in the enzyme transition state complex formed with each isomer. For a reaction involving two enantiomers, R and S, with specificity constants $(k_{cat}/K_m)_R$ and $(k_{cat}/K_m)_S$ respectively, the enantioselectivity E is related to $\Delta\Delta G^{\ddagger}$, the difference in free energy of the transition state complexes for R and S according to equation 5.

$$E = \frac{(k_{cat}/K_m)_R}{(k_{cat}/K_m)_S} = \exp(-\Delta\Delta G^{\ddagger}/RT) \qquad (5)$$

Thus enantioselectivity arises because enzyme and enantiomeric substrates form diastereomeric transition states of different energy $\Delta\Delta G^{\ddagger}$.

Although quantitative treatment of enzyme-catalysed chiral resolution is possible [99,100] if the value of E has been determined, it is rarely possible to predict the stereochemistry of catalysis for a given substrate unless an accurate active site model (such as that available for pig liver esterase [101]) or much comparative data on the selectivity for related substrate models is available. The effects of organic solvents on stereoselectivity are still more difficult to predict quantitatively. In a study of the stereoselectivity of trypsin, chymotrypsin, elastase, α-lytic protease and subtilisin for N-acetylalanine-chloroethyl ester, it was predicted that selectivity would decrease with increasing hydrophobicity of the solvent [102]. It was found that the ratio of specificity constants $(k_{cat}/K_m)_L/(k_{cat}/K_m)_D$ for L and D isomers decreased by 2 to 3 orders of magnitude when the reaction was performed in butyl ether rather than water. The data were interpreted in terms of differences in the ability of the substrate to displace water from the hydrophobic binding regions of the enzymes in aqueous and organic media. The effective reversal of selectivity in hydrophobic solvents has enabled peptides containing D-amino acid residues to be synthesized using subtilisin [103].

The maintenance [77,80,93] or enhancement [104,105] of enantioselectivity of lipase-catalysed reactions with chiral acids, alcohols or esters in organic solvents is well known. Lipases retain activity and selectivity in a wide range of solvents but some lipases show little activity or are denatured in aprotic, cyclic ethereal or chlorinated solvents [79]. Studies of direct esterification using lipases from Candida cylindracea [77] or porcine pancreas [106] showed a positive correlation between maximum rate of reaction and maximum enantioselectivity. A small amount of water or water mimicking solvent such as formamide or ethylene glycol was required for maximum activity and enantioselectivity. It was rationalised that multiple hydrogen-bonding in the enzymes active site was necessary for full enantioselectivity [106,107]. Improved selectivity of Candida cylindracea lipase has also been achieved by adding enantioselective inhibitors [108] or refolding the enzyme from organic solvent in the presence of an activator to form a stable conformer with high activity and selectivity [91].

Predictions of the stereochemical outcome of reactions catalysed by lipases in organic media would be particularly useful as the broad specificity of these enzymes has enabled their use for chiral resolution of a wide variety of molecules. The enantioselectivity of lipase-mediated transesterification depends on the complex interaction of both kinetic and thermodynamic factors but it is sometimes possible to make predictions as to the effects of solvent on selectivity which are useful in the design of enzymic chiral resolutions. For example, Chen and Sih [73] considered lipase-catalysed transfer of a chiral acyl moiety (X^R and X^S) from a chiral carboxylic acid ester to an achiral nucleophilic acceptor R^2OH (R^2OH may be an alcohol for the transesterification reaction or water for the hydrolytic reaction). The reaction mechanism may be represented by Figure 4.

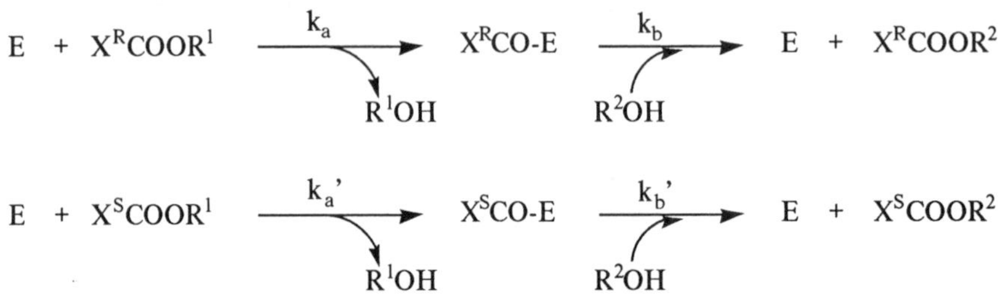

$$E + X^RCOOR^1 \xrightarrow{k_a} X^RCO\text{-}E \xrightarrow{k_b} E + X^RCOOR^2$$
$$\searrow R^1OH \qquad \nearrow R^2OH$$

$$E + X^SCOOR^1 \xrightarrow{k_a'} X^SCO\text{-}E \xrightarrow{k_b'} E + X^SCOOR^2$$
$$\searrow R^1OH \qquad \nearrow R^2OH$$

Figure 4. Reaction schemes for transesterification of R and S esters. k_a, k_a', k_b and k_b' are the net rate constants.

It was shown that if the reaction rate for the R and S enantiomers is V_R and V_S respectively, then:

$$\frac{Eo}{V^R} = \frac{1}{k_a[X^R]} + \left(\frac{1/k_b + k_a'[X^S]}{k_ak_b'[X^R]}\right) \cdot \frac{1}{[R^2OH]} \tag{6}$$

$$\frac{Eo}{V^S} = \frac{1}{k_a'[X^S]} + \left(\frac{1/k_b' + k_a[X^R]}{k_a'k_b[X^S]}\right) \cdot \frac{1}{[R^2OH]} \tag{7}$$

The selectivity of the transesterification reaction (in which the nucleophilic acceptor R^2OH is an alcohol) in organic media depends on the entire reaction sequence. However, for hydrolytic reactions in which R^2OH is both solvent and acceptor and is present at high concentration, equations 6 and 7 simplify to:

$$\frac{E_o}{V_R} = \frac{1}{k_a[X^R]} \tag{8}$$

and

$$\frac{E_o}{V_S} = \frac{1}{k_a'[X^S]} \tag{9}$$

Therefore, for hydrolytic reactions, selectivity is determined solely by the reaction steps leading to the formation of the acyl enzyme.

In this brief discussion, we have presented some examples of how enzyme conformation and the role of active-site bound water in catalytic activity may be profoundly influenced by organic media. Although the study of enzyme catalysis in organic media is, as yet, in it's infancy, it is apparent that the choice of solvent and substrate, presence or water or water-mimicking solvent, enantioselective inhibitors and temperature should enable the synthetic chemist to gain a degree of control over the chemical and stereochemical outcome of enzyme catalysis.

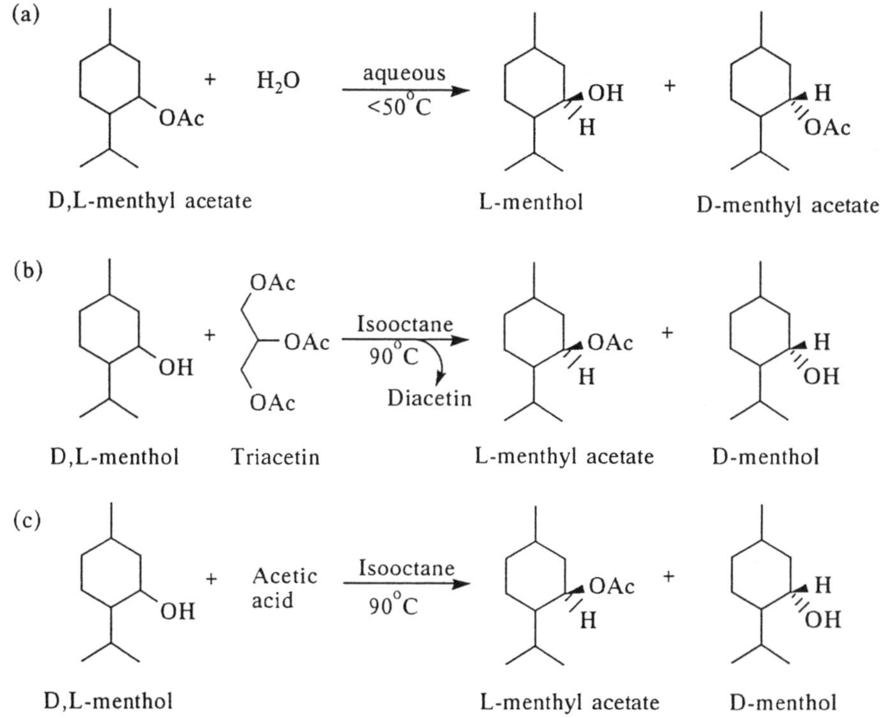

(a)

D,L-menthyl acetate   L-menthol   D-menthyl acetate

(b)

D,L-menthol   Triacetin   L-menthyl acetate   D-menthol

(c)

D,L-menthol   L-menthyl acetate   D-menthol

Figure 5. Lipase catalysed synthesis of D- and L-menthol stereoisomers.
(a) Hydrolysis of D,L-menthyl acetate; (b) Transesterification of D,L-menthol with triacetin and; (c) Esterification of D,L-menthol with acetic acid.

Examples of High Temperature Organic-phase Biocatalysis.

Example 1. Lokotch et al. [109] have compared the hydrolysis of D,L-menthyl acetate in water (Figure 5a) with the transesterification of D,L-menthol in triacetin (Figure 5b) and the direct esterification of D,L-menthol and acetic acid in isooctane (Figure 5c) using free and Lichrosorb-immobilized Candida cylindracea lipase.

The enzyme was stable in isooctane (little loss of activity was reported after 4.5 hours at 100°C) and transesterification was optimal at 90°C. In aqueous medium the free enzyme was denatured above 50°C but the hydrolytic rate was higher than that for transesterification or direct esterification (at 50°C transesterification activity was <30% of the hydrolytic rate at the same temperature). The thermal stability of the enzyme in organic solvent could be exploited to partially offset the relatively slow rates of transesterification by performing the reaction at high temperatures; e.g., transesterification activity at 90°C was <60% of the hydrolytic reaction at it's "optimal" temperature of 50°C.

Free enzyme was optimally active for transesterification with a water content of 6% but lyophilised enzyme (water content 3%) was inactive. Immobilised enzyme was optimally active at 2% water content and addition of excess water to the medium reduced significantly the conversion rate for both free and immobilised enzyme by promoting the back reaction. Water produced as a reaction product during direct esterification also promoted the back reaction. Transesterification was substrate

inhibited by triacetin and product inhibited by diacetin and it was concluded that
commercial exploitation would necessitate the development of a continuous fixed-bed
reactor process to minimise inhibition of the enzyme.

The enzyme was highly enantioselective for L-menthol and it's acetate in all
three types of reaction. Kinetic resolution via transesterification gave L-menthyl acetate
at 96% ee and D-menthol at 82% ee when the reaction was sampled at 48% overall
completion.

Example 2. Optically pure 2-substituted propionic acids are important as anti-
inflammatory agents and as intermediates in herbicide synthesis. Barton et al. [105]
have demonstrated a two-phase organic system using Candida cylindracea lipases to
resolve (R)-2-(4-hydroxyphenoxy)-propionic acid ((R)-HPPA) by hydrolysis of its
racemic methyl ester ((R,S)-HPPA-OMe). The enzyme showed good
enantioselectivity and catalytic rates in aqueous medium and in various
aqueous/hydrophobic solvent biphasic systems but was inactivated at 35°C or above.
In aqueous media the catalytic activity and enantioselectivity optima were pH 6 and pH
3 - 4 respectively while at pH values above 7, chemical hydrolysis resulted in
decreased enantiomeric purity of the product. The enzyme was product inhibited by
methanol and (R)-HPPA.

Under optimised conditions, adequate activity and enantioselectivity were
maintained in a batch reaction consisting of 25% toluene (the phase into which (R,S)-
HPPA-OMe partitioned) and 75% buffer, pH 6.5 (in which phase the products
predominated). Product inhibition could be alleviated by further diluting the aqueous
phase with water and in a kilogram scale reaction (R)-HPPA was produced at 88% ee
at 39% conversion.

Literature Cited

[1]   Laane, C. *Biocatalysis* 1987, *1*, 17.
[2]   Brink, L.E.S.; Tramper, *J. Biotechnol. Bioeng.* 1985, *27*, 1258.
[3]   Laane, C.; Boeren, S.; Vos, K. *Trends Biotechnol.* 1985, *3*, 251.
[4]   Halling, P.J. *Biotechnol. Bioeng.* 1990, *35*, 691.
[5]   Rekker, R.F. *The Hydrophobic Fragmental Constant,* Elsevier Publ., New
       York 1977.
[6]   Mozhaev, V.V., Khmelnitsky, Y.L., Sergeeva, M.V., Berlova, A.B.,
       Klyachko, M.L.; Levashov, A.V.; Martinek, K. *Eur. J. Biochem.* 1989, *184*,
       579.
[7]   Yamane, T. *Biocatalysis* 1988, *2*, 1.
[8]   Yamane, T.; Kojima, Y.; Ichiryu, T.; Shimizu, S. *Ann. N.Y.Acad. Sci.* 1988,
       542, 282.
[9]   Cassels, J.M.; Halling, P.J. *Enzyme Microb. Technol.* 1988, *10*, 486.
[10]   Dastoli, F.R.; Musto, N.A.; Price, S. *Arch. Biochem. Biophys.* 1966, *115*,
       44.
[11]   Klibanov, A.M. *Trends Briochem. Sci.* 1989, *14*, 141.
[12]   Wong, C-H.; Chen, S-T.; Hennen, W.J.; Bibbs, J.A.; Wang, Y-F.; Lui, J. L-
       C.; Pantoliano, M.W.; Whitlow, M.; Bryan, P.N. *J. Am. Chem. Soc.* 1990,
       112, 945.
[13]   Inada, Y.; Nishimura, H.; Takahashi, K.; Yoshimoto, T.; Saha, A.R.; Saito,
       Y. *Biochem. Biophys. Res. Comm.* 1984, *122*, 845.
[14]   Inada, Y.; Matsushima, A.; Takahashi, K.; Saito, Y. *Biocatalysis* 1990, *3*,
       317.
[15]   Sakurai, K.; Kashimoto, K.; Kodera, Y.; Inada, Y. *Biotechnol. Lett.* 1990,
       12, 685.

[16]   Brookes, I.K.; Lilly, M.D.; Drozd, J.W. *Enzyme Microb. Technol.* 1986, *8*, 53.
[17]   Owusu, R.K.; Cowan. D.A. *Enzyme Microb. Technol.* 1989, *11*, 568.
[18]   Davies and Rideal (1961) Interfacial Phenomena, Academic Press, NY, 474pp.
[19]   Jaycock and Parfitt (1981) Chemistry of Interfaces Ellis Harwood, 279pp.
[20]   Luisi, P.L. *Ang. Chem.* 1985, *24*, 439.
[21]   Luisi, P.L.; Giomini, M.; Pileni, M.P.; Robinson, B.H. *Biochim. Biophys. Acta.* 1988, *947*, 209.
[22]   Hilhorst, R.; Laane, C.; Veeger, C. *FEBS Lett.* 1983, *159*, 31.
[23]   Hilhorst, R.; Spruijt, R.; Laane, C.; Veeger, C. *Eur. J. Biochem.* 1984, *144*, 459.
[24]   Srivastava, R.C.; Madamwar, D.B.; Vyas, V.V. *Biotechnol. Bioeng.* 1987, *29*, 901.
[25]   Ruckenstein, E.; Karpe, P. *Biotechnol. Lett.* 1990, *4*, 241.
[26]   Khelnitsky, Y.L.; Zharinova, I.N.; Berezin, I.V.; Levashov, A.V.; Martinek, K. *Ann. N.Y. Acad. Sci.* 1987, *501*, 161.
[27]   Aaltonen, O.; Rantakylä *Chemtech.* 1991, *April*, 240.
[28]   Randolph, T.W.; Blanch, H.W.; Prausnitz, J.M. *Am. Institute Chem. Eng. J.* 1988, *34*, 1354.
[29]   Chi, Y.M.; Nakamura, K.; Yano, T. *Agric. Biol. Chem.* 1988, *6*, 1541.
[30]   Weaver, J.C.; Cooney, C.L.; Tannenbaum, S.R.; Fulton, S.P. In *Biomedical Applications of Immobilized Enzymes and Proteins* (Ming Swi Chang, ed.) Plenum Press, New York, 1977, pp. 207-255.
[31]   Olsen, L.F.; Degn, H. *Biochim. Biophys. Acta* 1978, *523*, 321.
[32]   Pulvin, S.; Legoy, M.D.; Lortie, R.; Pensa, M.; Thomas, D. *Biotech. Lett.* 1986, *8*, 783.
[33]   Hou, C.T. *Appl. Microb. Biotechnol.* 1984, *19*, 1.
[34]   Cedano, M.; Waissbluth, M. In *Enzyme Engineering* (Brown, C.B., Maneckee, G. and Wingard, L.B., eds.) Plenum Press, 1978, vol. 4, pp. 405-407.
[35]   Pulvin, S.; Sharan, M.; Thomas, D. *Anal. Biochem.* 1982, *124*, 258.
[36]   Barzana, E.; Klibanov, A.M.; Martinek, K. *Appl. Biochem. Biotechnol.* 1987, *15*, 25.
[37]   Barzana, E.; Karel, M.; Klibanov, A.M. *Biotechnol. Bioeng.* 1989, *34*, 1178.
[38]   Zaks, A.; Klibanov, A.M. *J. Biol. Chem.* 1988, *263*, 3194.
[39]   Zaks, A.; Klibanov, A.M. *Science* 1984, *224*, 1249.
[40]   Russel, A.J.; Klibanov, A.M. *J. Biol. Chem.* 1988, *263*, 11624
[41]   Singer, S.J. *Adv. Prot. Chem.* 1962, *17*, 1.
[42]   Herskovits, T.T.; Gadegbeku, B.; Jaillet, H. *J. Biol. Chem.* 1970, *245*, 2588.
[43]   Inoue, H.; Timasheff, S.N. *Biochemistry* 1968, *7*, 2501.
[44]   Izumi, T.; Yoshimusa, Y.; Inoue, H. *Arch. Biochem. Biophys.* 1980, *200*, 444.
[45]   Arakawa, T.; Godette, D. *Arch. Biochem. Biophys.* 1985, *240*, 21.
[46]   Schrier, E.E.; Ingwall, R.T.; Scheraga, H.A. *J. Phys. Chem.* 1965, *69*, 289.
[47]   Wuthrich, K. In *Protein Folding* (Jaenicke, R., ed.) Elsevier Press, 1980.
[48]   Dreyfus, M.; Vandenbunder, B.; Buc, H. *FEBS Lett.* 1978, *95*, 185.
[49]   Singh, T.J.; Wang, J.H. *J. Biol. Chem.* 1979, *254*, 8466.
[50]   Sakurai, H. *J. Biochem.* 1981, *90*, 95.
[51]   Khelnitsky, Y.L.; Levashov, A.V.; Klyachko, N.L.; Martinek, K. *Enzyme Microb. Technol.* 1988, *10*, 710.
[52]   Ingalls, R.G.; Squires, R.G.; Butler, L.G. *Biotechnol. Bioeng.* 1975, *17*, 1627.
[53]   Butler, L.G.; Reithel, F.J. *Arch. Biochem. Biophys.* 1977, *178*, 43.

[54]   Schwabe, C. *Biochemistry* 1969, *8*, 795.
[55]   Semenov, A.N.; Khmelnitski, Y.L.; Berezin, I.V.; Martinek, K. *Biocatalysis* 1987, *1*, 3.
[56]   Nakanishi, K.; Matsumo, R. *Ann. N.Y. Acad. Sci.* 1990, *613*, 652.
[57]   Adlercreutz, P.; Clapes, P.; Mattiasson, B. *Ann. N.Y. Acad. Sci.* 1990, *613*, 517
[58]   Cêrôvsky, V.; Martinek, K. *Coll. Czech. Chem. Commun.* 1984, *54*, 266.
[59]   Bevinakatti, H.S.; Banerji, A.A. *Biotechnol. Lett.* 1988, *10*, 397.
[60]   Hiratake, J.; Inagaki, M.; Nishioka, T.; Oda, J. *J. Org. Chem.* 1988, *53*, 6130.
[61]   Chapineau, J.; McCafferty, F.D.; Therisod, M.; Klibanov, A.M. *Biotechnol. Bioeng.* 1988, *31*, 208.
[62]   Miller, D.A.; Prausnitz, J.M.; Blanch, H.W. *Enzyme Microb. Technol.* 1991, *13*, 98.
[63]   Ergan, F.; Trani, M.; André, G. *Biotechnol. Lett.* 1988, *10*, 211.
[64]   Ottolina, G.; Carrea, G.; Riva, S. *J. Org. Chem.* 1990, *55*, 2336.
[65]   Ergan, F.; Trani, M.; André, G. *Biotechnol. Bioeng.* 1989, *35*, 195.
[66]   Hajjar, A.B.; Nicks, P.F.; Knowles, C.J. *Biotechnol. Lett.* 1990, *12*, 825.
[67]   Ghogare, A.; Kumar, G.S. *J. Chem. Soc. Chem. Commun.* 1989, 153.
[68]   Uemura, A.; Nozaki, K.; Yamashita, J.; Yasamuto, M. *Tetrahedron. Lett.* 1989, *30*, 3817.
[69]   Terao, Y.; Tsuji, K.; Murata, M.; Achiwa, K.; Nishio, T.; Watanabe, N.; Seto, K. *Chem. Pharm. Bull.* 1989, *37*, 1653.
[70]   Bianchi, D.; Cesti, P.; Battistel, E. *J. Org. Chem.* 1988, *53*, 5531.
[71]   Kullmann, W. *Enzymatic Peptide Synthesis*, CRC Press, Bocca, Raton, 1987, 140pp.
[72]   Perrin, D.D. *Aus. J. Chem.* 1964, *17*, 484.
[73]   Chen, C.-S.; Sih, C.J. *Angew. Chem. Int. (Ed. Eng.)* 1989, *28*, 695.
[74]   Koshland jr, D.E. *Proc. Natl. Acad. Sci.* 1958, *44*, 98.
[75]   Koshland jr, D.E. *J. Cell. Comp. Physiol.* 1959, *54*, 245.
[76]   Monod, J.; Wyman, J.; Changeux, J-P. *J. mol. Biol.* 1965, *12*, 88.
[77]   Gerlach, D.; Schreier, P. *Biocatalysis* 1989, *2*, 257.
[78]   Langrand, G.; Baratti, J.; Bhono, G.; Triantaphylides, C. *Tetrahedron Lett.* 1986, *27*, 29.
[79]   Hirata, H.; Higuchi, K.; Yamashina, T. *J. Biotechnol.* 1990, *14*, 157.
[80]   Dordick, J.S. *Enzyme Microb. Technol.* 1989, *11*, 194.
[81]   Kieboom, A.P.G. *Recl. Trav. Chim. Pays-Bas.* 1988, *107*, 347.
[82]   Zaks, A.; Russell, A.J. *J. Biotechnol.* 1988, *8*, 259.
[83]   Zaks, A.; Klibanov, A.M. *J. Am. Chem. Soc.* 1986, *108*, 2767.
[84]   Ferjansis, A.; Puigserver, A.; Gaertner, H. *Biotechnol. Lett.* 1988, *10*, 101.
[85]   Blakeney, A.B.; Stone, B.A. *FEBS Lett.* 1985, *186*, 229.
[86]   Pal, P.K.; Gertler, M.M. *Thromb. Res.* 1983, *29*, 175.
[87]   Kobayashi, M.; Nagasawa, T.; Yamada, H. *Appl. Microb. Biotechnol.* 1988, *29*, 231.
[88]   Ciuffreda, P.; Colombo, D.; Ronchetti, F.; Toma, L. *J. Org. Chem.* 1990, *55*, 4187.
[89]   Zemek, J.; Kucar, S.; Anderle, D. *Coll. Czech. Chem. commun.* 1988, *53*, 1851.
[90]   Ballesteros, A.; Bernabé, M.; Cruzado, C.; Martin-Lomas, M.; Otero, C. *Tetrahedron* 1989, *45*, 7077.
[91]   Wu, S.-H.; Guo, Z.-W.; Sih, C.J. *J. Am. Chem. Soc.* 1990, *112*, 1990.
[92]   Guo, Z.-W.; Sih, C.J. *J. Am. Chem. Soc.* 1988, *110*, 1999.
[93]   Wong, C.-H. *Science* 1989, *244*, 1145.
[94]   Colonna, S.; Gaggero, N.; Manfredi, A.; Casella, L.; Gullotti, M. *J. Chem. Soc. Chem. Comm.* 1981, *45*, 1451.

[95]   Baba, N.; Mimura, M.; Hiratake, J.; Uchida, K.; Oda, J. *Agric. Biol. Chem.*
       1988, *52*, 2685.
[96]   Rosario-Jensen, T.; Jiang, R.-T.; Tsai, M.-D. *Biochemistry* 1988, *27*, 4619.
[97]   Plant, A.R. *Pharm. Manufact. Rev.* 1991, *3*, 5.
[98]   Bednarski, M.D.; Simon, T.; Waldmann, H.; Whitesides, G.M. *J. Am. Chem.
       Soc.* 1989, *111*, 627.
[99]   Chen, C.-S.; Fujimoto, G.; Girdaukas, G.; Sih, C.J. *J. Am. Chem. Soc.*
       1982, *104*, 7294.
[100] Chen, C.-S.; Wu, S.-H.; Girdaukas, G.; Sih, C.J. *J. Am. Chem. Soc.* 1987,
       109, 2812.
[101] Toone, E.J.; Werth, M.J.; Jones, J.B. *J. Am. Chem. Soc.* 1990, *112*, 4946.
[102] Sakurai, T.; Margolin, A.L.; Russell, A.J.; Klibanov, A.M. *J. Am. Chem.
       Soc.* 1988, *110*, 7236.
[103] Margolin, A.L.; Tai, D.-F.; Klibanov, A.M. *J. Am. Chem. Soc.* 1987, *109*,
       7885.
[104] Yamamoto, K.; Nishioka, T.; Oda, J.; Yamamoto, Y. *Tetrahedron Lett.* 1988,
       29, 1717.
[105] Barton, M.J.; Hamman, J.P.; Fichter, K.; Calton, G.J. *Enzyme Microb.
       Technol.* 1990, *12*, 577.
[106] Kitaguchi, H.; Itoh, H.; Ono, M. *Chem. Lett.*, 1990, *1203*.
[107] Kitaguchi, H.; Klibanov, A.M. *J. Am. Chem. Soc.* 1989, *111*, 9272.
[108] Guo, Z.-W.; Sih, C.J. *J. Am. Chem. Soc.* 1989, *111*, 6836.
[109] Lokotsh, W.; Fritsche, K.; Syldatk, C. *Appl. Microbiol. Biotechnol.* 1989,
       31, 467.
[110] Luck, W.A.P. *Pure Appl. Chem.* 1987, *59*, 1215.
[111] Reslow, M.; Adlercreutz, P.; Mattiasson, B. *Ann. N.Y. Acad. Sci.* 1988,
       542, 250.

RECEIVED January 15, 1992

Chapter 8

Pressure Dependence of Enzyme Catalysis

Peter C. Michels and Douglas S. Clark

Department of Chemical Engineering, University of California—Berkeley, Berkeley, CA 94720

Pressure effects on enzyme function are strongly influenced by environmental factors such as temperature, pH, salt and substrate concentrations, and pressure itself. This article examines these effects in relation to volume changes associated with enzymatic reactions and enzyme denaturation. Depending on the experimental conditions, elevated pressure can enhance or inhibit enzyme activity and increase or decrease enzyme stability. Difficulties inherent in the design and interpretation of high pressure experiments are discussed, as are economic considerations pertaining to high pressure enzyme technology.

Intuitively, pressure effects on physicochemical processes are simple to understand as a direct extension of Le Chatelier's principle: elevated pressures favor changes that reduce a system's overall volume. Thus, by comparing the total volumes of the reactants versus products, of the ground state versus activated state, or of the dissociated versus bound complex, the effect of pressure on reaction equilibria or rates can be estimated. If the volume of the products is smaller than the volume of the reactants, the reaction equilibrium will shift to favor the products during a rise in pressure. Similarly, if the overall volume of the activated state is less than that of the ground state, the velocity of the reaction will increase with pressure.

The effect of pressure on the equilibrium constant of a reaction can be derived from the equilibrium condition:

$$\sum_i v_i \mu_i = 0 \qquad (1)$$

where v_i and μ_i are the stoichiometric coefficient and chemical potential of species i, respectively. Substituting the definition of activity into equation (1) yields

$$\sum_i v_i \mu_i^0 = -RT \ln \prod_i a_i^{v_i} = -RT \ln K_a \qquad (2)$$

where μ_i^0 is the chemical potential of species i at the standard state, defined here as one mole of pure component i at the system temperature and pressure, a_i is the

0097–6156/92/0498–0108$06.00/0

activity of species i, and K_a is the equilibrium constant defined in terms of activities. Equivalently, we can write

$$\sum_i \nu_i \mu_i^0 \ = \ -RT(\ln K_\gamma + \ln K_x) \tag{3}$$

where K_γ and K_x are the equilibrium constants defined in terms of activity coefficients and mole fractions, respectively. The pressure dependence of K_γ is described by

$$\left(\frac{\partial \ln K_\gamma}{\partial P}\right)_T = \sum_i \nu_i \left(\frac{\bar{v}_i - v_i^0}{RT}\right) \tag{4}$$

where $\bar{v}_i$ is the partial molar volume of species i in solution, and v_i^0 is the standard state molar volume of species i, defined by the basic thermodynamic relation

$$\left(\frac{\partial \mu_i^0}{\partial P}\right)_T = v_i^0 \tag{5}$$

Differentiating equation (3) with respect to pressure and combining the result with equations (4) and (5) gives the desired expression:

$$\Delta V = -RT\left(\frac{\partial \ln K_x}{\partial P}\right)_T \tag{6}$$

where ΔV is the total volume change for the system at a given pressure.

Similarly, the effect of pressure on rates of reaction can be derived from thermodynamics and transition state theory:

$$\Delta V^\# = -RT\left(\frac{\partial \ln k_x}{\partial P}\right)_T \tag{7}$$

where k_x is the rate constant based on pressure-insensitive molalities, and $\Delta V^\#$ is the classical activation volume. In physical terms, the classical activation volume represents the difference in partial molar volumes between the activated complex and the reactants.

In practice, ΔV and $\Delta V^\#$ are difficult to measure directly due to the many possible sources of volume changes in a reacting enzyme system. Interpreting measured pressure effects then entails the challenging task of accounting for all the contributions to the system volume, and determining which are the most significant. The volume of a reacting enzyme system can be divided into three basic components: 1) the volume of the enzyme itself, which includes the constitutive volume of the atoms, an incompressible void volume due to inefficient packing and imperfect folding, and a compressible void volume; 2) the volumes of all substrates and products, as well as cofactors and inhibitors involved in the reaction; and 3) the volume of the solvent. Changes in any of these volumes during a reaction will manifest a pressure effect on the process. Other responses to pressure, such as shifts in the pKa of protein side groups and buffers (1-4), alterations in the structure of water (5-6), and conformational shifts of the protein can also be attributed to changes in one or more of these three volume components. These effects, in turn, largely account for the overall behavior of proteins under pressure, including association or dissociation of subunits, denaturation or unfolding, changes in binding affinity for ligands, and modified catalytic rates. Still other effects of pressure can stem from changes in the viscosity of the reaction medium (3), altered solubility of reactants (7), and pressure-sensitive fluidity and permeability of membranes (8-9). These factors must be recognized for an accurate evaluation of volume changes during enzyme reactions to be made.

The behavior of enzymes under pressure is further complicated by the wide range of experimental parameters that can influence pressure effects. Experimental conditions such as temperature, pH, ionic strength, ionic composition, and ligand concentrations can be critical, altering not only the magnitude but even the sign of ΔV's . Considering this, ΔV values compiled in the literature (7,10,11) must be viewed very carefully, and comparison of two values from different reports may have little meaning. From an engineering perspective, optimization of enzyme properties under pressure should examine the dependence of volume changes on all relevant variables. Thus, a clear understanding of the interactive effects of pressure and other environmental factors is fundamental to conducting properly controlled and complete studies. In this review, we briefly describe the influence of pressure on various aspects of enzyme catalysis, including ΔV's associated with enzyme reactions.

Pressure Effects on Enzyme Activity and Stability

The effects of pressure on enzyme activity are conveniently described by an overall activation volume, ΔV^*:

$$\Delta V^* = RT \frac{\ln(v_1/v_2)}{(P_2 - P_1)} \tag{8}$$

where v_1 is the reaction rate at P_1. The overall activation volume does not have the same physical significance as the classical activation volume of a single chemical reaction [equation (7)] but is still useful for quantifying pressure effects on enzymatic rates. Alternatively, the dependence of k_{cat} on pressure can be described by $\Delta V_{k_{cat}}$:

$$\Delta V_{k_{cat}} = -RT \left(\frac{\partial \ln k_{cat}}{\partial P} \right)_T \tag{9}$$

Overall activation volumes (and in some cases $\Delta V_{k_{cat}}$ values) for a wide range of enzymatic reactions and conditions have been determined and are compiled in several recent reviews (7,12,13). Few general trends emerge from these collections; for instance, Penniston (14) has suggested that monomeric enzymes are activated by high hydrostatic pressure whereas multimeric proteins are inhibited. However, several counter-examples have been reported, including at least fifteen dimeric and tetrameric enzymes that are activated by pressure (7). Among the more striking examples is the hydrogenase activity in crude extracts of the deep-sea thermophile *Methanococcus jannaschii*, which increased more than three-fold as the pressure was increased from 7.5 to 260 atm (85). This rate enhancement corresponds to an overall activation volume of ca. -140 cc/mol. In addition, Morild's extensive compilation includes 31 negative activation volumes, and 70 values which are either zero or positive. As discussed further below, these values can be affected dramatically by the conditions of the reaction, so neither positive nor negative activation volumes seem to be predominant. Typically, however, ΔV^* ranges from ca. +60 cc/mol to -70 cc/mol, with most values having a magnitude of less than 30 cc/mol.

Favorable effects of pressure on rates can only be exploited if the active enzyme conformation is stable at elevated pressures. In general, the effect of pressure on protein denaturation is described by ΔV_d, the volume change associated with unfolding:

$$\Delta V_d = RT \left(\frac{\partial \ln K_d}{\partial P} \right)_T \qquad (10)$$

where K_d is the equilibrium coefficient describing the ratio of active to inactive conformations of the enzyme. Published values of ΔV_d suggest that the pressure sensitivity of proteins relates to their quaternary structure: typical values of ΔV_d's are ca. -30 to -100 cc/mol for the most stable monomeric proteins (11,15), ca. -100 cc/mol for the dissociation of multimeric proteins (16), and up to ca. -10,000 cc/mol for the degradation of very large protein assemblies (17). Interestingly, these values would indicate that proteins are destabilized by pressure, and that rate enhancements under pressure must be short-lived. However, most of these determinations of ΔV_d have been made by observing pressure-induced denaturation at pressures of several kilobars--beyond the range where life is currently known to exist. At more moderate pressures characteristic of deep-sea habitats, there is substantial evidence that ΔV_d can be positive. Apparent stabilization of proteins in this "physiological" pressure range has a long history in the literature (17-21), and only some of the more recent examples will be reviewed here.

Suzuki and Tanaguchi (22) describe two proteins (β -lactoglobulin and α -amylase) that show enhanced renaturation of the pressure denatured form when pressures of 1000-3000 atm are applied, and two proteins (γ -globulin and ovalbumin) which exhibit improved resistance to thermal denaturation at pressures between 1000-4000 atm. Morita and Haight (23) demonstrated that malic dehydrogenase from the thermophile *Bacillus stearothermophilus* was stabilized above the boiling point of water by 1300 atm of pressure. In a similar study (24), the upper temperature limit of pyrophosphatase activity increased from below 100°C at atmospheric pressure to above 105°C at pressures between 1300 and 1700 atm. In addition, the hydrolysis of pyrophosphate catalyzed by pyrophosphatase proceeded with a 25-fold increase in yield at 700 atm compared to the same temperature at atmospheric pressure. In a separate study (25), Morita found that aspartase from *Escherichia coli* was both stabilized and activated at moderately elevated temperatures (45 - 56°C) by pressures up to 1000 atm. Similarly, thermal inactivation of malate dehydrogenase was retarded by pressures up to 1000 atm (26). More recently, the thermal half-life of amyloglucosidase was increased by 400% through the application of just 200 atm (27). Thus, pressure-enhancements of both enzymatic stability and activity are common within the biologically relevant range of pressures.

To summarize, the contention that proteins are destabilized by very high pressures (typically above 3000 atm) appears undebatable. However, the extent to which proteins are stabilized by low to moderate pressures must still be considered on a case-by-case basis. The evidence needed to establish stabilization includes increased thermal half-lives under pressure and/or improved renaturation induced by pressure.

Temperature Effects on ΔV Values

In addition to the trend towards higher temperature stability at higher pressures, increased barostability has been observed at temperatures slightly above the habitat temperature from which the organism (or enzyme) was isolated (10,28-29). Higher temperatures have also been found to favorably decrease the activation volume for some enzyme reactions (30-32), and representative results are included in Table 1. Magnification of pressure enhancements [e.g., chymotrypsin in Table 1 (33)], as well as reduction of pressure inhibition of catalytic rates [e.g., pyruvate kinase (34)] have both been observed. Favorable influences are not universal, however, as evidenced by alkaline phosphatase at low salt concentration (35) (Table 1). While the underlying causes of these observations are not entirely clear, examining the effect of temperature on the intermolecular forces which govern protein volume changes is informative.

Elevated temperatures are known to disrupt the highly ordered structure of electrostricted water (36-37). Electrostriction, the strong contraction of hydration volume due to alignment of dipolar water molecules in the electric field of exposed charge, often produces the most dramatic volume changes in enzymatic reactions. By extrapolating the data of Parsons (38), Nickerson concluded that electrostriction should become less significant at higher temperatures and may be eliminated entirely at 90-115°C (39). Minimizing electrostriction at high temperatures reduces the strongly destabilizing effect of pressure on ion pairs, which are important for the stability and activity of many proteins (40-42). Indeed, Nickerson hypothesizes that the interdependent effects of high temperature, high salt concentration, and high pressure may be the stabilizing factors responsible for allowing inherently unstable peptides and biological macromolecules to evolve on the early Earth (39).

Despite much speculation, the effect of pressure on hydrophobic interactions and their role in protein denaturation is poorly understood. Since hydrophobic amino acids tend to concentrate near the interior of proteins, denaturation is expected to bring nonpolar side chains into contact with water. Unfortunately, due to the difficulty of adequately modeling the complex behavior of proteins, experimental information on volume changes associated with such transitions are scarce and contradictory (43-46). Some studies suggest that volume changes for the exposure to water of buried hydrophobic groups are positive due to structural changes of the surrounding water (46). However, high temperatures will inhibit the formation of highly hydrogen-bonded water cages (sometimes referred to as "icebergs") around hydrophobic groups and thus reduce any accompanying volume increase. This effect would diminish the positive contribution to the ΔV of unfolding and tend to offset the advantage of reduced electrostriction promoted by high temperatures.

Pressure and pH Effects

As with temperature, pH changes can have a profound effect on the stability of folded proteins at atmospheric pressure. At elevated pressures, these effects are compounded, since pH affects not only K_d but also the electrostrictive contribution of ionizable groups to ΔV_d. Zipp and Kauzmann clearly illustrated these points using sperm whale myoglobin (47). Changes in pH dramatically shifted the P-T phase diagram as well as the maximum pressure at which the enzyme remained folded. Related data have been collected for other proteins (Table 2).

pH can also influence the sign and magnitude of volume changes for enzyme reactions. For example, the apparent activation volume for fructose diphosphatase decreased from zero to -40 cc/mole in response to a pH shift of one unit (48). In view of this result and the sensitivity of enzyme reactions to pH at atmospheric pressure, it is clear that great care should be taken to control pH in high pressure experiments. When designing such experiments, it must be recognized that the pKa's of many common buffers, acids, and bases are dramatically affected by pressure (1-4).

Pressure and Salt Effects

As with pH and temperature, salts can cause dramatic, and sometimes unpredictable, changes to ΔV's (Table 3). For instance, salts can raise (49-51), decrease (35, 51-52), or have no significant effect (35, 51-53) on the activation volume. In addition, ions have been shown to greatly enhance the barostability of organisms (54), multimers (55), and monomeric proteins (1). Furthermore, salt effects can vary with temperature (35).

Like temperature, salts can affect the hydration layer and modify volume changes associated with conformational shifts near the protein surface during catalysis or unfolding. The organization of water in the hydration layer can be altered by simple electrostatic effects (56, 57), which are solely dependent on the ionic strength (for example, pyruvate kinase/KCl in Table 3 (58)), and/or by more complex salting in/salting out phenomena (59-62), which depend on both the concentration and composition of the salt (as illustrated most dramatically by malate dehydrogenase in Table 3).

Table I. Temperature Effects on ΔV Parameters of Enzyme Reactions

Enzyme	Temperature (^{o}C)	ΔV_d (cc/mol)	ΔV^* (cc/mol)	Reference
Chymotrypsin	20	-72	-4.4	33
	30	-44		
	40		-6.5	
	45		-13	
	50		-35	
Metmyoglobin	5	-114		47
	40	-51		
Invertase	30		-8	17
	40		-69	
Phosphofructokinase	3		-11	81
(rattail)	28		-46	
Pyruvate kinase	10		+53	34
(squid)	25		+32	
Alkaline phosphatase	10		-27.5	35
	40		-14	
	10 (in 1M KI)		-1.0	
	20 (in 1M KI)		-8.5	

Table II. Effects of pH on ΔV Parameters of Enzyme Reactions

Enzyme	pH	ΔV_d (cc/mol)	ΔV^* (cc/mol)	Reference
Metmyoglobin	10	-60		47
	5	-95		
Ribonuclease	2.0	-38		82
	4.0	-16		
Fructose diphosphatase	7.7		0	48
	8.75		-40	

Table III. Salt Effects on ΔV^* Values of Enzyme Reactions

Enzyme	Salt	Concentration (mM)	ΔV^* (cc/mol)	Reference
Malate	KF	200	-27	51
dehydrogenase	K_2SO_4	200	-13	
(pig heart)	KCl	200	-2	
	KBr	200	0	
	KI	200	+20	
	KSCN	200	+23	
Lactate	KF	100	-11	51
dehydrogenase	K_2SO_4	100	-2	
(rabbit)	KCl	100	+6	
	KBr	100	+13	
	KI	100	+16	
	KSCN	100	+20	
Pyruvate kinase	K_2SO_4	200	+7	51
(*Scorpaena gutatta*)	KCl	200	+14	
	KBr	200	+25	
	KI	200	+37	
	KSCN	200	+54	
	KCl	0.01	+54	58
	KCl	10	+35	
	KCl	40	+17	
	KCl	80	+10.5	
	NH_4Cl	80	+7.6	
	NaCl	80	+54.3	
	LiCl	80	+61	
Lysozyme	NaCl	4	-9.7	49
	NaCl	125	+1.5	
Thermolysin	NaBr	20	-63	53
	$CaCl_2$	10	-63	
	$CaCl_2$	100	-63	
	$CaCl_2$	1000	-64	
	KCl	3000	-65	
Alkaline	KI	1	-27	35
phosphatase (10°)	KI	1000	-1	
Alkaline	KI	1	-21	35
phosphatase (20°)	KI	1000	-8.5	

Pressure Effects on Substrate Binding

Pressure effects on substrate binding are a primary cause of pressure sensitivity for many enzymatic reactions (16). This topic has been reviewed fairly extensively (1, 11, 12, 16, 63), so only a brief overview of binding interactions supplemented with recent data is provided here.

The effect of pressure on the binding equilibrium of substrates is described by ΔV_{K_s}, defined as

$$\Delta V_{K_s} = RT \left(\frac{\partial \ln K_s}{\partial P} \right)_T \tag{11}$$

Rigorously, this "binding volume" is the difference between the partial molar volumes of the unbound species and the bound enzyme-substrate complex; however, these values are difficult to evaluate directly. Usually, therefore, the pressure dependence of the Michaelis constant is determined and used to obtain the so-called reaction volume, ΔV_{K_m} (64-67):

$$\Delta V_{K_m} = RT \left(\frac{\partial \ln K_m}{\partial P} \right)_T \tag{12}$$

which contains contributions from the volume change of substrate binding. Once again, the effects of pressure vary widely. Ligand binding can be enhanced (1, 68), inhibited (16, 69), or unaffected (67, 70) by pressure, and selected examples appear in Table 4. The nature of both the enzyme binding site and the ligand are important. For example, the binding of substrates and inhibitors to the active site of thermolysin is significantly favored by elevated pressure, whereas binding to the homologous protease from *Bacillus subtilis* is slightly hindered (67). Similarly, the binding of proflavin to trypsin is diminished at high pressures, while binding to a similar protease, chymotrypsin, is unaffected by elevated pressure (70). Moreover, the binding of other ligands to chymotrypsin can be either hindered or enhanced (71) by pressure. Model studies by Taniguchi (72) and Le Noble et al. (73), as well as extensive kinetic investigations by Kunugi and co-workers (64, 67-69) and others (74-75), further demonstrate that ΔV_{K_m} for an enzyme can vary dramatically, and even change sign, depending on the substrate.

Economic Considerations

Thus far, most high pressure studies have been concerned with the mechanisms of adaptation of deep-sea life or with using pressure as a tool to elucidate the energetics and mechanisms of well characterized enzymes. However, the many examples of pressure-stabilized and pressure-activated enzymes, as well as increased under-

Table IV. Substrate Effects on ΔV Parameters of Enzyme Reactions

Enzyme	Substrate	ΔV^* (cc/mol)	Ref.
Horseradish	ferrocyanide	+6.7	65
peroxidase	aminobenzoic acid	-2.5	
	l-tyrosine	-9.4	
Chymotrypsin	2-valeryloxybenzoic acid	-20	66
	Suc-Ala-pNA	-30	
	Suc-Ala-Ala-pNA	-33	
Catalase	H_2O_2 (10 mM)	-5	50
	H_2O_2 (20 mM)	-24	
	H_2O_2 (39 mM)	0	
Malate	malate (5 mM)	+14	26
dehydrogenase	malate (20 mM)	+10	
	malate (75 mM)	+1	
Trypsin	Bz-Arg-pNA	-10	83
	Bz-Arg-OEt	-2.4	
	Bz-Arg-NH2	+6	

		ΔV^* (cc/mol)	ΔV_{K_m} (cc/mol)	Ref.
Carboxypeptidase Y	Fua-Gly-Phe	+27	+16	64
	Fua-Phe-Phe	+10	+22	
	Fua-Phe-OEt	-1.5	+6	
	Fua-Phe-OMe	-3.7	+6	
	Fua-Phe-NH2		+8	
Carboxypeptidase W	Fua-Phe-Gly		+27	69
	Fua-Phe-OEt		+3	
Carboxypeptidase A[a]	Phe-Phe		+34	84
	l-phenylalanine		+23	
	d-phenylalanine		+4	

[a] ΔV_{K_i} for inhibitor binding

Note: The sign convention is described in the text and may differ from the sign convention used in individual references.

standing of how to control and optimize pressure effects by changing the reaction environment, suggest that some industrial enzymatic processes might be improved by the application of high pressure.

Such improvements can only be realized at the expense of the higher capital and operating costs associated with high pressure chemical processes. No estimation of these economic considerations for biological reactions has yet been offered, and considering the wide range of variables presented by both biological and high pressure systems, only the most basic order-of-magnitude analysis is warranted at this time. As discussed by Isaacs (76) for high pressure, liquid phase organic syntheses, an activation volume of at least -30 cc/mol is desirable for a high pressure process. Although relatively few enzymatic reactions exhibit this large a pressure enhancement, the additional benefits of increased catalyst stability (27), improved selectivity and yield (77), unique specificities (78), reduced product inhibition (79), and novel recovery schemes (80) should, in some cases at least, increase the feasibility of high pressure biotechnology. In addition, further opportunities for exploiting the benefits of elevated pressure should parallel the continuing discovery of novel organisms and enzymes from deep-sea and terrestrial high-pressure habitats.

Acknowledgments

The authors are grateful for the financial support of the Office of Naval Research (N00014-89-J-1884) and the National Science Foundation (PYI Award of D.S. Clark). P.C. Michels is a National Science Foundation Fellow.

Literature Cited

1.   Jaenicke, R. In *Current Perspectives in High Pressure Biology*; Jannasch, H.;Marquis,R. E.; Zimmerman, A. M., Eds.; Academic Press Inc.: London, 1987; 257-273.
2.   Rasper, J.; Kauzmann, W. *J. Am. Chem. Soc.*, **1962**, *84*, 1771-1777.
3.   Kauzmann, W.; Bodansky, A.; Rasper, J. *J.Am. Chem. Soc.*, **1962**, *84*, 1777-1788.
4.   Neuman, R. C.; Kauzmann, W.; Zipp, A. *J. Phys. Chem.*, **1973**, *77*, 2687-2691.
5.   Suzuki, K.; Taniguchi, Y.;Tsuchiya, M. In *High Pressure Science and Technology*; Timmerhaus, H. D.; Barber, M. S., Eds; Proceedings of the Sixth AIRAPT Conference; Plenum Press: New York, NY, **1977**, 548-554.
6.   Francks, F *Water: A Complete Trestise*; Plenum Press: New York, NY, **1972**; Vol. 1.
7.   Morild, E. *Adv. Pro. Chem.*, **1981**, *34*, 93-166.
8.   Chong, P. L. G.; Fortes, P. A. G.; Jameson, D. M. *J. Biol. Chem.*, **1985**, *260*, 14484-14490.
9.   Gibbs, A.; Somaro, G. N. *J.Exp.Biol.*, **1989**, *143*, 475-492.
10. Jaenicke, R. *Ann. Rev. Biophys. Bioeng.*, **1981**, *10*, 1-67.
11. Heremans, K. *Ann. Rev. Biophys. Bioeng.*, **1982**, *11*, 1-21

12. Isaacs, N. S. *Liquid Phase High Pressure Chemistry*, John Wiley and Sons: London, **1981**.
13. Ludlow, J. M.; Clark, D. S. *Crit. Rev. Biotech.*, **1991**, *10*, 321-345.
14. Penniston, J. T. *Arch. Biochem. Biophys.*, **1971**, *142*, 322-332.
15. Taniguchi, Y. In *High Pressure Chemical Synthesis*; Jurczak, J.; Baranowski, B., Eds.; Elsevier: Amsterdam, **1989**; 349-373.
16. Siebenaller, J. F.; Somaro, G. N. *CRC Crit. Rev. Aqua.Sci.*, **1989**, *1*, 1-25.
17. Johnson, F. H.; Eyring, H.; Polissar, M. J. *The Kinetic Basis of Molecular Biology*; John Wiley & Sons: New York, NY, **1954**.
18. Bresler, S. E. *Compt. Rend. Acad. Sci*, **1947**, *55*, 141-147.
19. Johnson, F. H.; Campbell, D. H. *J. Biol. Chem.*, **1946**, *163*, 689-701.
20. Suzuki, K *Rev. Phys. Chem. Jap.*, **1962**, *29*, 86-102.
21. Johnson, F. H.; Eyring, H. In *High Pressure Effects on Cellular Processes*; Zimmerman, A. M., Ed. ; Academic Press: London, 1970; 2-44.
22. Suzuki, K.; Taniguchi, Y. *Symp. Soc. Exp. Biol.*, **1972**, *26*, 103-124.
23. Morita, R. Y.; Haight, R. D. *J. Bacteriol.*, **1962**, *83*, 1341-1346.
24. Morita, R. Y.; Mathemeier, P. F. *J. Bacteriol.*, **1964**, *88*, 1667-1671.
25. Haight, R. D.; Morita, R. Y. *J. Bacteriol.*, **1962**, *83*, 112-120.
26. Berger, L. R. In *Proceedings of the Fourth International Conference on High Pressure*, Osugi,J. Ed.; Physico-chemical Society of Japan: Kyoto, **1974**, 639-643.
27. Rohrbach, R. P.; Mallarik, M. J. *United States Patent 4,415,656*, **1983**.
28. Johnson, F. H. *Symp.Soc. Gen. Microbiol.*, **1957**, *7*, 134-167.
29. Yayanos, A. A. *Proc.Natl. Acad. Sci. USA*, **1986**, *83*, 9542-9546.
30. Landau, J. V. *Science*, **1966**, *153*, 1273-1274.
31. Fitos, I.; Heremans, K. *React. Catal. Lett.*, **1979**, *12*, 399-403.
32. Mustafa, T.; Moon, T. W.; Hotchatchka, P. W.; *Am. Zool.*, **1971**, *11*, 451-466.
33. Tanaguchi, Y.; Suzuki, K. *J. Phys. Chem*, **1983**, *87*, 5185-5193.
34. Storey, K. B.; Hotchatchka, P. W. *Comp. Biochem. Physiol. B*, **1975**, *52*, 187-191.
35. Greaney, G. S.; Somaro, G. N. *Biochemistry*, **1979**, *18*, 5322-5332.
36. Whalley, E. *J. Chem. Phys.*, **1963**, *38*, 1400-1405.
37. Hamann, S. D. *Rev. Phys. Chem. Jap.*, **1980**, *50*, 147-169.
38. Parsons, R. *Handbook of Electrochemical Constants*; Butterworths: London, **1959**.
39. Nickerson, K. W. *J. Theor. Biol.*, **1984**, *110*, 487-499.
40. Perutz, M. F. *Science*, **1978**, *201*, 1187-1191.
41. Hocking, J. D.; Harris, J. I. *Eur. J. Biochem.*, **1980**, *108*, 567-579.
42. Walker, J. E. *FEBS Symp.*, **1979**, *52*, 211-225.
43. Kauzmann, W. *Nature*, **1987**, *325*, 763-764.
44. Nemethy, G.; Scheraga, H. A. *J. Phys. Chem.*, **1962**, *66*, 1773-1789.
45. Klapper, M. H. *Biochim. Biophys. Acta*, **1971**, *229*, 557-566.
46. Hvidt, A. ; *J. Theor. Biol.*, **1975**, *50*, 245-252.

47. Zipp, A.; Kauzmann, W. *Biochemistry*, **1973**, *12*, 4217-4228.
48. Hotchatchka, P. W.; Behrisch, H. W.; Marcus, F. *Am. Zool.*, **1971**, *11*, 437-449.
49. Neville, W. M.; Eyring, H. *Proc. Natl. Acad. Sci. USA*, **1972**, *69*, 2417-2419.
50. Morild, E.; Oldheim, J. E. *Physiol. Chem. & Phys.*, **1981**, *13*, 419-428.
51. Low, P. S.; Somaro, G. N. *Proc.Nat.Acad.Sci. USA*, **1975**, *72*, 3014-3018.
52. Johnson, F. H.; Kauzmann, W.; Gensler, R. L. *Arch. Biochem.*, **1948**, *19*, 229-236.
53. Fukuda, M.; Kunugi, S. *Biocatalysis*, **1989**, *2*, 225-233.
54. Palmer, D. S.; Albright, L. J. *Liminol. Oceanog.*, **1970**, *15*, 343-347.
55. Schade, B. C.; Ludemann, H. -D.; Rudolph, R.; Jaenicke, R. *Biochemistry*, **1980**, *19*, 1121-1130.
56. Millaro, F. J. *J. Phys. Chem.*, **1969**, *73*, 2417-2420.
57. Wirth, H. E. *J. Am. Chem. Soc.*, **1948**, *70*, 462-465.
58. Low, P. S.; Somaro, G. N. *Comp. Biochem. Physiol.*, **1975**, *52B*, 67-74.
59. Long, F. A.; McDevit, W. F. *Chem. Rev*, **1952**, *51*, 119-132.
60. Von Hippel, P. H.; Shleich, T. In *Structure and Stability of Biological Macromolecules*; Timasheff, S. N.; Fasman, G. D., Eds.; Marcel-Dekker: New York, NY, **1969**, 417-574.
61. Robinson, D. R.; Jencks, W. P.; *J. Am. Chem. Soc.*, **1965**, *87*, 2470-2477.
62. Melander, W.; Horvath, C. *Arch. Biochem. Biophys.*, **1977**, *183*, 200-215.
63. Weber, G.; Drickamer, H. G. *Quar. Rev. Biophys.*, **1983**, *16*, 89-112.
64. Fukuda, M.; Kunugi, S. *Eur. J. Biochem.*, **1985**, *149*, 657-662.
65. Ralston, I. M.; Wauters, J.; Heremans, K. *Biophys. Chem.*, **1982**, *15*, 15-18.
66. Makimoto, S.; Taniguchi, Y. *Biochim. Biophys. Acta*, **1987**, *914*, 304-307.
67. Fukuda, M.; Kunugi, S. *Eur. J. Biochem.*, **1984**, *142*, 565-570.
68. Muller, K.; Ludemann, H. D.; Jaenicke, R. *Biophys. Chem.*, **1981**, *14*, 101-110.
69. Fukuda, M.; Kunugi, S. *J. Biochem.*, **1987**, *101*, 233-240.
70. Heremans, K.; Snauwaert, J.; Vanderspayn, H.; Van Nuland, Y. In *Proceedings of the Fourth International Conference on High Pressure*, Osugi,J. Ed.; Physico-chemical Society of Japan: Kyoto, **1974**, 623-626.
71. Werbin, H.; McLaren, A. D. *Arch. Biochim.*, **1951**, *31*, 285-288.
72. Taniguchi, Y.; Makimoto, S.; Suzuki, S. *J. Phys. Chem.*, **1981**, *85*, 3469-3472.
73. le Noble, W. J.; Srivastava, S.; Breslow, R.; Trainor, G. *J. Am. Chem. Soc.*, **1983**, *105*, 2745-2747.
74. Makimoto, S.; Suzuki, S.; Taniguchi, Y. *J. Phys. Chem.*, **1984**, *88*, 6021-6024.
75. Makimoto, S.; Suzuki, S.; Taniguchi, Y. *Bull. Chem. Soc.Jpn.*, **1986**, *59*, 243-247
76. Isaacs, N. S. In *High Pressure Chemical Synthesis*; Jurczak, J.; Baranowski, B., Eds.; Elsevier: Amsterdam, **1989**; 349-373.
77. Kunugi, S.; Tanabe, K.; Yamashita, K.; Morikawa, Y.; Ito, T.; Kondoh, T.; Hirata, K.; Nomura, A. *Bull. Chem. Soc. Jpn.*, **1989**, *62*, 514-518.
78. Kunugi, S.; Tanabe, K.; Fukuda, M.; Taniguchi, Y. *J. Chem. Soc., Chem. Commun.*, **1987**, 1335-1336.
79. Morild, E. *Biophys. Chem.*, **1977**, *6*, 351-362.

80. Hayashi, R.; Kawamura, Y.; Kunugi, S. *J. Food Sci.*, **1987**, *52*, 1107-1108.
81. Moon, T. W.; Mustafa, T.; Hotchatchka, P.W. *Am. Zool.*, **1971**, *11*, 467-472.
82. Brandts, J. F.; Oliviera, R. J.; Westort, C. *Biochemistry*, **1970**, *9*, 1038-1047.
83. Fukuda, M.; Kunugi, S.; Ise, N. *Biophys. Biochim. Acta*, **1982**, *704*, 107-113.
84. Fukuda, M.; Kunugi, S.; Ise, N. *Bull. Chem. Soc. Jpn.*, **1983**, *56*, 3308-3313.
85. Miller, J. F.; Nelson, C. M.; Shah, N. N.; Clark, D.S. *Biotechnol. Bioeng.*, **1989**, *34*, 1015-1021.

RECEIVED January 15, 1992

Chapter 9

Thermodynamic Strategies for Protein Design
Increased Temperature Stability

Martin Straume, Kenneth P. Murphy, and Ernesto Freire

Department of Biology and Biocalorimetry Center, The Johns Hopkins University, Baltimore, MD 21218

Different thermodynamic strategies aimed at increasing the temperature stability of proteins are discussed. The properties of the "average globular protein" have been used as a starting point to assess the effects of specific interactions to thermostability. Changes in the hydrogen bond content, hydrophobic interactions, configurational entropy of the unfolded state, the presence of specific ligand binding sites, and protein size have been considered. The results of this analysis provide a guide to evaluate the magnitude of the stabilizing effects of these interactions.

The identification of organisms exhibiting optimal growth temperatures as high as 105°C (1-3) has provided a source for naturally occurring proteins that maintain their functional and structural properties to temperatures well beyond those at which most mesophilic proteins have lost all structural integrity (4-6). The availability of proteins from such thermophilic organisms has offered unique experimental systems with which to address questions regarding the origin of their high stability with the hope of finding specific features that can later be used in the design or modification of other proteins.

Efforts directed toward identifying the basis of protein thermostability in thermophilic organisms have, to date, largely focussed on comparisons of amino acid compositions and sequences between those proteins and their mesophilic counterparts. This analysis has suggested somewhat larger proportions of hydrophobic residues in thermophilic proteins, some specific amino acid substitutions (e.g. Gly->Ala) in several proteins, higher proportions of proline residues, and higher proportions of the hydrophobic residues leucine, isoleucine and phenylalanine. However, in those instances in which these changes have been observed, they have been relatively small on a percent basis. For example, the glyceraldehyde-3-phosphate dehydrogenases isolated from four related microorganisms exhibiting three different optimal growth temperatures, T_{opt}, of 37°C (*Methanobacterium bryantii* and *Methanobacterium formicicum*), 83°C (*Methanothermus fervidus*), or 100°C (*Pyrococcus woesii*), show increases in the total apolar surface area, but only on the order of approximately 1-4% (7,8). Net increases in alanine content of 4-8 residues and reductions of glycine of 0-8 residues have also been observed, but not all as direct amino acid substitutions (7,8). In fact, many globular proteins from mesophilic organisms have higher proportions of some thermostabilizing structural

0097–6156/92/0498–0122$06.00/0

elements than many thermophilic proteins. This is demonstrated by the average of approximately 76 ± 5 Å^2 of apolar surface per amino acid observed in a variety of proteins from thermophilic organisms (*9-14*) compared to an average of approximately 86 Å^2 of apolar surface area per amino acid observed in the mesophilic proteins considered by Murphy and Freire (*15*). A detailed analysis of the role of hydrogen bond content must await specific three-dimensional structural information before accurate comparisons become possible.

The conclusion that has emerged from studies of the amino acid composition and structural features of various thermostable proteins is that proteins derived from thermophilic sources are not endowed with any particularly unique structural properties that are responsible for their enhanced thermostability. Instead, thermophilic proteins appear to derive their enhanced thermostability by combinations of the same protein structure-stabilizing interactions that impart structural stability to proteins from mesophiles, i.e. hydrophobic interactions, hydrogen bonds, van der Waals interactions, ionic interactions, ligand binding, disulfide linkages, etc.

The contribution of numerous diverse types of stabilizing effects therefore requires simultaneous consideration of a variety of energetic influences on the overall stability of protein structures. Such a multifaceted analytical problem is a challenging one due to the fact that protein structures are only marginally stable (i.e. the free energy of stabilization is typically of the order of 5 - 20 kcal mol^{-1}). The overall structural stability of proteins is comprised of a very large number of largely compensatory energetic terms that ultimately add up to produce a rather small overall free energy of stabilization. The secret of thermophilic proteins appears to be the alteration of the final count by subtle structural modifications.

The purpose of this chapter is to present a detailed quantitative analysis of the individual effects of the most important structural interactions that contribute to the thermal stability of a protein. The properties of an "average" mesophilic globular protein will be used as a starting point and the effects of specific structural modifications on its stability will be considered. These specific modifications include the degree of hydrogen bonding, hydrophobic interactions, configurational entropy, ligand interactions, and molecular weight.

Thermal Stability of Globular Proteins

In order to study quantitatively the structural stability of proteins, it is necessary to determine the magnitude of the free energy change, $\Delta G°$, associated with the overall process of folding a protein from its denatured state into its native conformation. Since the temperature dependence of protein stability is of interest, it is also necessary to know the thermodynamic terms which specify the temperature dependence of the free energy change, namely the enthalpy change, $\Delta H°$, the entropy change, $\Delta S°$, and the heat capacity change, $\Delta C_p°$. At any temperature, the Gibbs free energy change can be written in terms of the standard thermodynamic relationship:

$$\Delta G° = \Delta H°(T_R) + \Delta C_p° (T - T_R) - T (\Delta S°(T_R) + \Delta C_p° \ln(T / T_R)) \qquad (1)$$

where T_R is an appropriate reference temperature. In this and all equations in this chapter the native state of the protein will be used as the reference state.

Several interactions contribute to the overall magnitude of the parameters in equation 1. They include primarily hydrophobic interactions, hydrogen bonding, van der Waals interactions, electrostatic interactions, and configurational entropy changes. Additionally, interactions with specific ligands, metal ions, urea, protons, etc. must be characterized for those cases in which these additional interactions are present.

The magnitude of the thermodynamic parameters that determine the stability of globular proteins has been extensively studied using modern high sensitivity calorimetric techniques (16). Analysis of the existing protein database with supplemental information provided by model compound studies has permitted a quantitative dissection of the forces that control protein stability. In this chapter, we will summarize those results; for a complete discussion the reader is referred to the original literature (15-18).

It was noted as early as 1974 (6) that both $\Delta H°$ and $\Delta S°$, when normalized with respect to the number of amino acids in the protein, converge to common values at characteristic temperatures. The temperatures at which they converge are designated T_H* and T_S* and the values at these temperatures are designated $\Delta H*$ and $\Delta S*$ (17). For globular proteins, the convergence temperatures T_H* and T_S* are close to 100°C and 112°C, respectively (18). Based on model compound results, it has been argued that at T_H* and T_S*, the hydrophobic contributions to the enthalpy and entropy changes are zero (18). In the case of the model compounds, the dissolution of homologous series, in which the polar group was unchanged but the amount of apolar surface was varied, was studied. These systems also show convergence temperatures for $\Delta H°$ and $\Delta S°$. It can be mathematically demonstrated for these series of compounds that at the convergence temperature, the apolar contribution to $\Delta H°$ or $\Delta S°$ of dissolution is zero and that the convergence value, $\Delta H*$ or $\Delta S*$, corresponds to the polar $\Delta H°$ or the mixing $\Delta S°$ (18).

The protein denaturation data shows convergence when normalized to the number of residues (or the molecular weight). It was reasoned (18) that the protein data shows convergence because globular proteins have a relatively constant number of hydrogen bonds per residue (16), but a variable number of buried apolar groups per residue so that, when normalized to the number of residues, they constitute a homologous series of compounds.

Consequently, the existence of convergence temperatures allows a separation of the hydrophobic or apolar contributions to the overall energetics of protein stabilization. Accordingly, the apolar contributions to the enthalpy and entropy changes are given by:

$$\Delta H_{ap} = \Delta C_{p,ap} (T - T_H*) \tag{2}$$

$$\Delta S_{ap} = \Delta C_{p,ap} \ln(T / T_S*) \tag{3}$$

where $\Delta C_{p,ap}$ is the apolar contribution to the overall $\Delta C_p°$. $\Delta C_{p,ap}$ is proportional to the apolar surface area (ASA) that becomes exposed to the solvent upon protein unfolding and can be expressed per Å^2 exposed (0.44 cal (K mol Å^2)$^{-1}$) or equivalently in terms of the number of apolar hydrogens (i.e. hydrogens bound to carbon) that become exposed to the solvent upon protein unfolding (6.69 cal (K mol aH)$^{-1}$) (18). Analysis of structural data for globular proteins indicates that the fraction of buried apolar hydrogens is proportional to the size of the protein and can be expressed as:

$$f_{aH} = 0.6142 + 0.0006273 N_{res} \tag{4}$$

where N_{res} is the number of amino acid residues in the protein (18). It follows that $\Delta C_{p,ap}$ can be written as:

$$\Delta C_{p,ap} = N_{aH} f_{aH} \Delta C_{p,aH} \tag{5}$$

where $\Delta C_{p,aH}$ is the contribution to $\Delta C_{p,ap}$ per apolar hydrogen that becomes exposed to the solvent upon unfolding and N_{aH} is the total number of apolar hydrogens in the protein.

By definition, ΔH^* and ΔS^* contain all other contributions to the energetics of protein stability. Under standard conditions (neutral pH, absence of ligand binding effects, etc.) ΔH^* is equal to 1.35 ± 0.03 kcal $(\text{mol-res})^{-1}$ and ΔS^* is equal to 4.3 ± 0.12 cal $(\text{K mol-res})^{-1}$ (*18*). It has been argued (*15,18*) that under standard conditions ΔH^* contains essentially the effective contributions arising from hydrogen bonding (i.e. the contributions arising from the overall process of breaking hydrogen bonds within the protein, the associated disruption of van der Waals interactions, and the formation of hydrogen bonds and van der Waals interactions with water). Since, on average, proteins have 0.72 hydrogen bonds per amino acid residue (*16*), the effective enthalpy change per hydrogen bond can be estimated to be on the order of *1.87* kcal $(\text{mol H-bond})^{-1}$ at T_H^*.

The denaturational entropy change contains terms for the entropy change in the solvent, due to the exposure of apolar and polar groups, and a term for the gain in configurational entropy of the polypeptide chain, given as the product of the number of residues and ΔS^*. Under standard conditions, the polar contributions to the enthalpy and entropy changes can be written as:

$$\Delta H_{pol} = N_{hb} \left[\Delta H_{hb} + \Delta C_{p,hb} (T - T_H^*) \right] \tag{6}$$

$$\Delta S_{pol} = \Delta C_{p,pol} \ln(T / T_S^*) \tag{7}$$

where N_{hb} is the number of hydrogen bonds, ΔH_{hb} the enthalpy change per hydrogen bond, and $\Delta C_{p,hb}$ is the heat capacity change for the exposure of a hydrogen bond pair. $\Delta C_{p,hb}$ has been estimated to be -14.3 cal $(\text{K mol H-bond})^{-1}$ using the cyclic dipeptide data (*18*).

The overall free energy change for denaturing the protein is the sum of all the constituent effects:

$$\Delta G^\circ = \Delta G_{ap} + \Delta G_{pol} + \Delta G_{conf} + \Delta G_{other} \tag{8}$$

where ΔG_{conf} is the configurational free energy change and ΔG_{other} includes any specific effects such as ligand binding. These equations have successfully been used to model both the stability and the cooperative unfolding of myoglobin (*19*) and phosphoglycerate kinase (*20*).

Strategies for Engineering Temperature Stability

The Average Globular Protein. Using the thermodynamic parameters and equations discussed above it is possible to describe the stability of an "average" single domain globular protein. According to the protein database for which accurate structural and thermodynamic data exist (*15*), that protein will have approximately 150 amino acid residues, a total of 856 apolar hydrogens, 108 hydrogen bonds and will bury about 606 apolar hydrogens. The predicted temperature stability of such a protein is shown in figure 1. In panel A, the free energy of stabilization is shown. This quantity displays a characteristic curvature and crosses zero twice. The two temperatures at which the free energy is zero define the cold and heat denaturation temperatures respectively. Under standard conditions, the cold denaturation temperature is well below 0°C and cannot be observed experimentally. However under certain conditions (e.g. low pH, the presence of denaturants) this temperature can be increased to an accessible experimental range (*22*). The temperature at which ΔG° is maximal defines the temperature at which the stability of the protein is maximal.

Panel B in figure 1 shows the predicted heat capacity function expected for that protein. This function is the one measured by differential scanning calorimetry

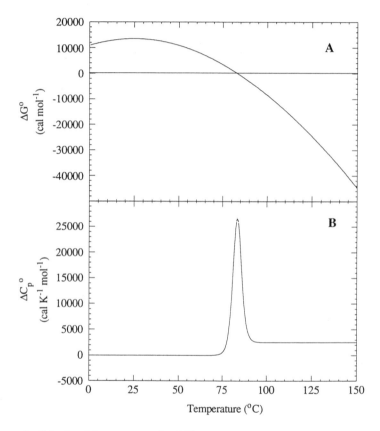

Figure 1. Calculated free energy of stabilization (A) and heat capacity function (B) for an "average" globular protein (150 amino acids, 856 apolar hydrogens, 108 hydrogen bonds, and 606 buried apolar hydrogens). The thermodynamic parameters and equations presented in the text were used to generate these predicted free energy and heat capacity profiles. A T_m of 83.5°C results with a transition enthalpy of 160 kcal mol^{-1} and a heat capacity change of 2.5 kcal (K mol)$^{-1}$.

and contains all the information for a thermodynamic characterization of the folding/unfolding transition (23,24). In calculating the heat capacity function, a two-state transition mechanism was assumed, as is the case for the globular proteins discussed here (5,16,23-25). The heat denaturation transition in panel B is characterized by a transition temperature (T_m) of 83.5°C, a transition enthalpy change of 160 kcal mol^{-1} and a heat capacity change of 2.5 kcal (K mol)$^{-1}$. As expected, these values represent average values for globular proteins (16).

According to the above discussion, there are three generic structural parameters that largely determine the thermal stability of a globular protein under standard conditions: (a) the buried apolar surface area, (b) the number of hydrogen bonds, and (c) the number of residues in the protein. From the point of view of developing rational methods for protein design and protein modification, it is necessary to estimate the expected changes in protein stability elicited by variations in the above parameters. This can be done using the formalism discussed here.

Effects of Buried Apolar Surface Area. Changes in the buried apolar surface area, or equivalently in the number of buried apolar hydrogens, will change the magnitude of the hydrophobic contributions to protein stability. On the average, the mesophilic globular proteins for which accurate structural and thermodynamic data exist (15) have a total of 5.7 apolar hydrogens per amino acid. The variation found in globular proteins ranges from 5.36 for ribonuclease A to 6.39 for myoglobin. Of the globular proteins studied, these are the ones that exhibit the lowest and highest $\Delta C_p°$ values upon denaturation. The number of apolar hydrogens per amino acid varies from two for glycine to ten for leucine and isoleucine. From an engineering point of view, however, a more significant quantity is the ratio of the number of apolar hydrogens to the total side chain area, since the design of specific amino acid substitutions requires that the packing of the amino acid side chains be taken into account.

Figure 2 shows the effect of increasing the number of apolar hydrogens per residue on the stability of the average globular protein. For these calculations the number of apolar hydrogens per residue was increased from 5.7 to 7.3 while keeping all the other contributions constant. As shown in panel A, increasing the hydrophobic stabilization has two effects on the Gibbs free energy curves, it increases their absolute magnitude while simultaneously shifting the maximum to higher temperatures. The immediate consequence of this mechanism of stabilization is an increase in the Tm for both the cold and heat denaturation temperatures. Thus, while the protein becomes more stable at higher temperatures it becomes less stable at low temperatures. Panel B shows the expected heat capacity versus temperature curves. These curves are centered at progressively higher temperatures and display increasing $\Delta C_p°$ values. The increase in $\Delta C_p°$ is the characteristic signature for this type of effect and reflects the exposure to water of a larger number of apolar surfaces as the protein unfolds.

Effects of Hydrogen Bonding. As discussed above, globular proteins have an average of 0.72 hydrogen bonds per residue which for a protein of 150 amino acids is equivalent to 108 hydrogen bonds. The effect of increasing the number of hydrogen bonds on protein stability while keeping all the other interactions constant is shown in Figure 3. Panel A shows the Gibbs free energy curves obtained upon increasing the number of hydrogen bonds per residue from 0.72 to 0.84 and panel B the corresponding heat capacity curves. The curve with the highest T_m in the figure corresponds to an additional 18 hydrogen bonds in the protein and represents a stabilization of almost 50°C.

According to the above calculations, the addition of a single hydrogen bond to the average globular protein contributes an additional 1.5 kcal mol^{-1} to the Gibbs free

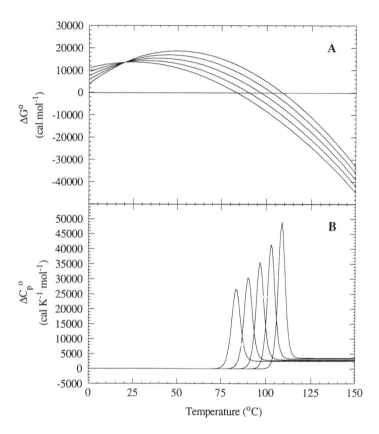

Figure 2. Changes in free energies of stabilization (A) and heat capacity functions (B) resulting from increasing the number of apolar hydrogens per amino acid from 5.7 (the average) to 7.3 while keeping all other contributions constant. ΔG_{max} increases in magnitude and in temperature producing both higher heat and cold denaturation temperatures. $\Delta C_{p,max}$ also increases in magnitude and in temperature and the ΔC_p^o for the transition also increases as more apolar surfaces are exposed to water upon unfolding.

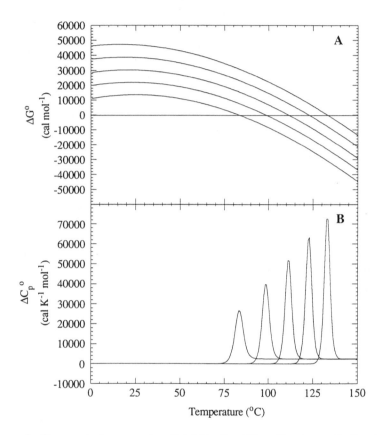

Figure 3. Changes in free energies of stabilization (A) and heat capacity functions (B) resulting from increasing the number of hydrogen bonds per amino acid from 0.72 (the average) to 0.84 while keeping all other contributions constant. The value of ΔG_{max} increases substantially producing a stabilization of almost 50°C in the T_m, as seen also in the increasing temperature of $\Delta C_{p,max}$.

energy of stabilization at the transition temperature. This additional contribution results in an increase in T_m of 3.5°C. In order to achieve a similar increase in T_m by hydrophobic interactions, an additional 20 apolar hydrogens would need to be buried in the interior of the protein.

Effects of Protein Size. The number of amino acid residues has an indirect effect on the stability of a globular protein due to the geometrical relationship between the total protein volume and its surface. This effect results in an increase in the fraction of apolar hydrogens packed within the protein (see equation 4) and indirectly in an increase in the relative magnitude of the hydrophobic interactions. Since, for globular proteins, the number of hydrogen bonds (and hence the free energy contribution from hydrogen bonds) is found to be linearly related to the number of amino acids, this effect by itself does not contribute to the size dependence of T_m. The same is true for the configurational entropy change which varies linearly with the number of residues.

Figure 4 shows the effect of increasing the number of amino acids in the protein from 150 to 270. While this effect is relatively small, the Tm increases from 83.5°C to 94°C upon doubling the molecular weight, it nevertheless provides a vehicle for stabilization especially if this size increase can be used to change the hydrogen bonding or hydrophobic composition of the protein.

Effects of Configurational Entropy. For globular proteins, the average value of the configurational entropy change is $4.3 + 0.12$ cal $(K \text{ mol-res})^{-1}$. An increase in this value has a destabilizing effect on the protein while a decrease in this value has a thermostabilizing effect. Within this context, the so called increased "rigidity" of the native state of thermophilic proteins will tend to increase the magnitude of the entropy change and contribute to destabilize the structure. The increased rigidity might be the price paid by the existence of a larger number of stabilizing interactions in the protein. One way of reducing the entropy change and stabilizing the protein is by restricting the configurational freedom of the denatured state. In addition to the obvious restrictions placed by the presence of covalent interactions like disulfide bridges, other effects are possible. For example, within the native conformation, amino acid side chains are in rather rigid configurations independently of their size; upon unfolding the configurational entropy gain of a residue depends on its side chain. Studies of the a-helix to coil transition have indicated that the entropy change varies among different amino acids (26). Also, mutation studies on T4 lysozyme have indicated that Gly->Ala substitutions decrease the entropy gain of the unfolded state resulting in protein stabilization. Theoretically, the decrease in entropy gain has been estimated to be -2.4 cal $(K \text{ mol})^{-1}$ for a single Gly->Ala substitution (27). Experimentally, a single Gly->Ala substitution was found to contribute on the order of 1 kcal mol^{-1} to the free energy of stabilization, resulting in an increase of ~1°C in the transition temperature of T4 lysozyme (28). Substitution in favor of proline residues is also consistent with this effect (28).

Figure 5 shows the effects of decreasing the average unfolding entropy change per amino acid from 4.3 to 3.98 cal $(K \text{ mol-res})^{-1}$. Even though the effects shown in the figure are large, such large entropic changes will require drastic changes in amino acid composition. For example, while Gly->Ala substitutions have been observed in thermophilic proteins, these substitutions by themselves are not enough to account for a significant proportion of the enhanced thermostability. As discussed before the strategy in nature seems to have been that of a large number of small effects that together result in a significant stabilization of the protein structure.

Effects of Specific Ligands. Another mechanism of protein stabilization is through the formation of specific ligand, metal ion binding sites or prosthetic groups

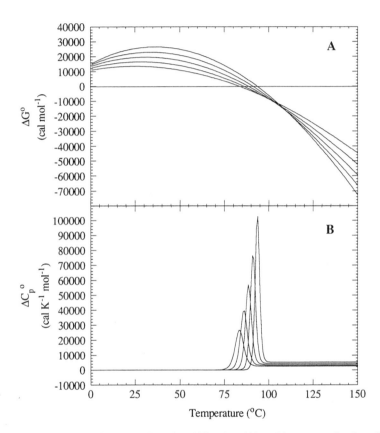

Figure 4. Changes in free energies of stabilization (A) and heat capacity functions (B) resulting from increasing the size of the protein (i.e. the number of amino acid residues) from 150 to 270 amino acids while keeping all other contributions constant. The value of ΔG_{max} is seen to increase and to shift to higher temperatures, however, the T_m increase is relatively small even after doubling the size of the protein.

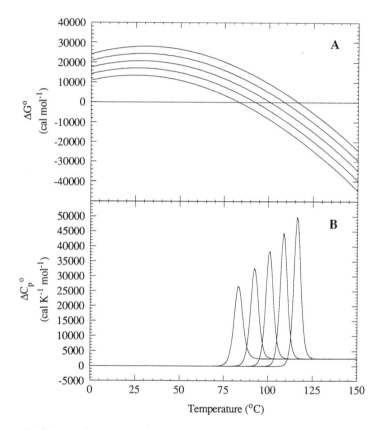

Figure 5. Changes in free energies of stabilization (A) and heat capacity functions (B) resulting from decreasing the average unfolding entropy change per residue from 4.3 cal (mol residue)$^{-1}$ K^{-1} (the average) to 3.98 cal (mol residue)$^{-1}$ K^{-1} while keeping all other contributions constant. The ΔG° profile is shifted upward thus increasing the T_m to higher temperature. Although the effects demonstrated here are quite substantial, drastic changes in amino acid composition would have to occur to elicit such a large decrease in the average unfolding entropy per residue.

in the native conformation of the protein. In this case, the overall Gibbs free energy of stabilization will contain an additional term of the form:

$$\Delta G = \Delta G° + nRT \ln(1 + K_b[X]) \qquad (9)$$

$\Delta G°$ is the Gibbs free energy of stabilization in the absence of the ligand molecule, n is the number of binding sites, K_b the binding constant of the ligand to the protein and [X] the free concentration of ligand. In this case, the stability of the protein explicitly depends on the concentration of ligand, and as such it can be the subject of environmental regulation. Also, amino acid changes might result in enhanced binding affinities. Figure 6 shows the effects of ligand concentration on the stability of a protein possessing a single specific binding site. As shown in the figure, increasing the concentration of ligand to 104 times the magnitude of the dissociation constant $(1/K_b)$ or equivalently increasing the binding affinity by four orders of magnitude increases the Tm from 83.5°C to 97°C. An increase of this magnitude in the binding affinity is equivalent to an additional binding free energy of -5.5 kcal mol^{-1}.

Binding of specific ligands or the presence of prosthetic groups is known to be a rather general mechanism for stabilization of native protein structures. Numerous examples of these effects have been discussed in the literature with regard to both mesophilic and thermophilic proteins (*29,30*). Metal ions, Ca^{2+}, and Zn^{2+} are known to stabilize specific structural motifs in proteins (*31*). The structure of heme proteins is stabilized to a large extent by the presence of the heme group. For example, apo myoglobin is extremely unstable and apo cytochrome c is unable to fold in the absence of the heme group. An assessment of the role of ligand binding effects in naturally occurring thermophiles requires evaluation of affinity constants and free ligand concentration levels in thermophiles and their mesophilic counterparts.

Other Effects. Other more specific interactions, like the presence of salt bridges or disulfide bridges are also likely to contribute to thermostability. However, these effects require more specific protein modifications than the ones discussed above. While the presence of disulfide bridges has been acknowledged to play a significant role in the stabilization of several mesophilic proteins, there has been some discussion regarding a similar role in thermophilic proteins and particularly in extreme thermophiles. It has been argued that the enhanced tendency to oxidation at high temperatures will transform disulfide bridges into destabilizing factors under those conditions. This is consistent with the observation of a reduced cysteine content in thermophilic proteins relative to their mesophilic counterparts (*14*).

Concluding Remarks

The calculations presented here, in conjunction with the existing compositional and structural data for thermophilic proteins, indicate that the increased thermostability of those proteins cannot be accounted for by an enhancement of any particular interaction but in terms of the combined and cumulative effects of changes in most, if not all, interactions.

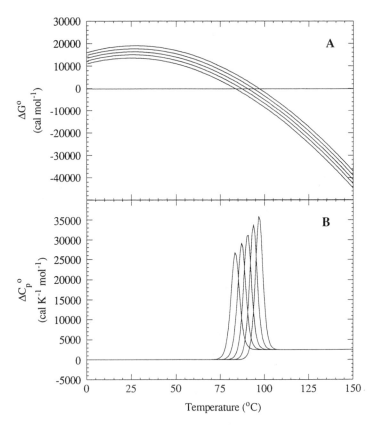

Figure 6. Changes in free energies of stabilization (A) and heat capacity functions (B) resulting from increasing the binding of a specific ligand to the native conformation of an "average" globular protein while keeping all other contributions constant. The ΔG_{max} profile moves upward with increasing concentration of ligand (to a level 10^4 times greater than the ligand dissociation constant) or equivalently by increasing the binding affinity by 10^4 (i.e. an additional -5.5 kcal mol^{-1} in the binding free energy). The magnitude of ΔG_{max} is increased as is the T_m as demonstrated by the higher temperature location of the $\Delta C_{p,max}$.

Acknowledgments

This work was supported by grants from the National Institutes of Health (RR04328, NS24520, and GM37911). KPM was supported by a postdoctoral fellowship from the National Science Foundation (DIR-8721059). All correspondence should be addressed to Ernesto Freire, Department of Biology and Biocalorimetry Center, The Johns Hopkins University, Baltimore, MD 21218.

Literature Cited

1   Stetter, K. O.; Lauerer, G.; Thomm, M.; Neuner, A. *Science*, **1987**, *236*, 822-824.
2   Fiala, G.; Stetter, K. O. *Arch. Microbiol.*, **1986**, *145*, 56-61.
3   Stetter, K. O. *Nature*, **1982**, *300*, 258-260.
4   Privalov, P. L. *Adv. Prot. Chem.*, **1982**, *35*, 1-104.
5   Privalov, P. L. *Adv. Prot. Chem.*, **1979**, *33*, 167-241.
6   Privalov, P. L.; Khnechinashvili, N. N. *J. Mol. Biol.*, **1974**, *86*, 665-684.
7   Zwickl, P.; Fabry, S.; Bogedain, C.; Haas, A.; Hensel, R. *J. Bacteriol.*, **1990**, *172*, 4329-4338.
8   Fabry, S.; Lang, J.; Niermann, T.; Vingron, M.; Hensel, R. *Eur. J. Biochem.*, **1989**, *179*, 405-413.
9   Rossi, M.; Rella, R.; Pisani, F.; Cubellis, M. V.; Moracci, M.; Nucci, R.; Vaccaro, C. In *Life Under Extreme Conditions*; diPrisco, G., Ed.; Springer-Verlag: Berlin, 1991; pp 115-123.
10 Wrba, A; Schweiger, A; Schultes, V.; Jaenicke, R. Biochemistry, 1990, 29, 7584-7592.
11 Aono, S.; Bryant, F. O.; Adams, M. W. W. *J. Bacteriol.*, **1989**, *171*, 3433-3439.
12 Bryant, F. O.; Adams, M. W. W. *J. Biol. Chem.*, **1989**, *264*, 5070-5079.
13 Schultes, V.; Deutzmann, R.; Jaenicke, R. *Eur. J. Biochem.*, **1990**, *192*, 25-31.
14 Hocking, J. D.; Harris, J. I. *Eur. J. Biochem.*, **1980**, *108*, 567-579.
15 Murphy, K. P.; Freire, E. *Adv. Prot. Chem.*, **1991**, in press.
16 Privalov, P. L.; Gill, S. J. *Adv. Prot. Chem.*, **1988**, *39*, 191-234.
17 Murphy, K. P.; Gill, S. J. *J. Mol. Biol.*, **1991**, in press.
18 Murphy, K.P.; Privalov, P. L.; Gill, S. J. *Science*, **1990**, *246*, 559-561.
19 Freire, E.; Murphy, K. P.; *J. Mol. Biol.*, **1991**, in press.
20 Freire, E.; Murphy, K. P.;Sanchez-Ruiz, J. M.; Galisteo, M. L.; Privalov, P. L.; *Biochemistry*, **1991**, in press.
21 Baldwin, R. L. *Proc. Natl. Acad. Sci. USA*, **1986**, *83*, 8069-8072.
22 Privalov, P. L.; Griko, Y. V.; Venyaminov, S. Y.; Kutyshenko, V. P. *J. Mol. Biol.*, **1986**, *190*, 487-498.
23 Freire, E.; Biltonen, R. L. *CRC Crit. Rev. Biochem.*, **1978**, *5*, 85-124.
24 Freire, E.; Biltonen, R. L. *Biopolymers*, **1978**, *17*, 481-494.
25 Freire, E. *Comments Mol. Cell. Biophys.*, **1989**, *6*, 123-140.
26 Sueki, M.; Lee, S.; Powers, S. P.; Denton, J. B.; Konishi, Y.; Scheraga, H. A. *Macromolecules*, **1984**, *17*, 148-155.
27 Nemethy, G.; Leach, S. J.; Scheraga, H. A. *J. Phys. Chem.*, **1966**, *70*, 998-1004.
28 Matthews, B. W.; Nicholson, H.; Becktel, W. J. *Proc. Natl. Acad. Sci. USA*, **1987**, *84*, 6663-6667.
29 Fontana, A. In *Life Under Extreme Conditions*; diPrisco, G., Ed.; Springer-Verlag: Berlin, 1991; pp 89-113.
30 Goins, B.; Freire, E. *Biochemistry*, **1988**, *27*, 2046-2052.
31 Branden, C.; Tooze, J. *Introduction to Protein Structure*; Garland Publishing, Inc.: New York, NY, 1991.

RECEIVED January 15, 1992

Chapter 10

Stability of High-Temperature Enzymes
Development and Testing of a New Predictive Model

Bruce E. Dale and John P. McBennett

Engineering Biosciences Research Center and Department of Chemical Engineering, Texas A&M University, College Station, TX 77843

Proteins are being considered or designed for operation in many unusual environments: high temperatures, extremes of pH, organic media, high salt levels, etc. As yet, there are no quantitative, fundamental models available to predict how particular proteins might respond to particular environments. One key protein property is stability, the thermodynamic tendency of the protein to maintain or reassume the folded form. Although many factors stabilize proteins, it is widely believed that hydrophobic effects are the dominant contributor to stability. We develop here a phase equilibrium approach to protein stability which models the protein molecule as a hydrophobic "phase" and the thermodynamic parameters of unfolding (enthalpy and heat capacity change) as average properties of the amino acids composing this protein phase. This model successfully predicts experimentally observed protein stability in several systems. This phase equilibrium approach to protein stability requires no sequence information or solved X-ray structures; only the amino acid composition of independently folding domains is required. Behavior of high temperature enzymes is predicted and interpreted in terms of the model.

Background

It is difficult to overstate the importance of protein stability in science, medicine and industry. However, reliable methods are not available for accurately predicting protein stability. The most rigorous definition of protein stability is in terms of the free energy change of unfolding under defined conditions. Computational techniques based on peptide interactions have been developed to estimate the conformational energy of proteins. Unfortunately, due to: a) the complexity of the conformational hyperspace even for small proteins, b) the large number of potential peptide interactions, and c) solvent effects; it is not yet possible to calculate conformational energy (and to thereby estimate protein stability) with any precision (1).

We develop here a phase equilibrium model for protein stability which views: a) the protein molecule as a hydrophobic phase in an aqueous environment and b) the unfolding process as a phase transition in which the amino acids are transferred from the hydrophobic protein "phase" to the surrounding aqueous phase (2). This model

0097–6156/92/0498–0136$06.00/0

is computationally simple and robust and provides many testable predictions of protein behavior.

There are at least three lines of evidence that suggest that such a phase equilibrium model for protein stability might prove fruitful: a) Unfolding of protein domains is an extremely cooperative process (*3*). This is characteristic of other phase transitions such as melting (hence the description of the temperature of protein unfolding as the "melting" temperature). b) The thermodynamics of dissolution of simple liquid hydrocarbons and solid cyclic dipeptides in water are qualitatively and quantitatively similar to those of proteins (*4,5*). c) A statistical mechanics thermodynamic model based on condensation (i.e., collapse of protein molecules into a phase) successfully predicts many of the important experimentally observed features of protein folding (*6*). Since many of the arguments for the appropriateness of a phase equilibrium model are based on the concept of the hydrophobic effect, the evidence that the hydrophobic effect is indeed a dominant contributor to protein stability is briefly reviewed.

Hydrophobicity: The Dominant Force in Protein Stability.   The following arguments that hydrophobicity is the dominant force in protein stability were recently summarized in an excellent review article by Dr. Ken A. Dill (*6*). First, there is a strong resemblance between the temperature dependence of the free energy of folding and the temperature dependence of the free energy of transfer of nonpolar model compounds from water into nonpolar media. Second, nonpolar residues are sequestered to the interior of the protein molecule where they largely avoid contact with water, in effect forming a hydrophobic phase in water. Third, protein stability is affected by different salt concentrations in the same order as the Hofmeister series. Fourth, replacement of a given residue by other amino acids shows that the stability of the resulting protein is proportional to the partitioning of the amino acid between a hydrophobic phase and water. Fifth, hydrophobicities of residues within the core of proteins are strongly conserved. Sixth, computer simulations of protein folding show that the interior/exterior distribution of hydrophobic residues is a key factor in incorrectly folded proteins. Therefore, there is ample justification for attempting to predict protein stability based on a model of the protein as a hydrophobic phase in water.

Development of a Phase Equilibrium Model for Protein Stability.

To recapitulate, if we accept the hypothesis that hydrophobicity is the dominant force governing protein stability then we can use this concept to develop a model for protein stability based on hydrophobicity. This approach (or model) treats the protein molecule as a phase and each amino acid as a component of the phase. Amino acids are modeled as contributing to a macroscopic protein "phase" property, such as the free energy change of unfolding, according to the number and type of each amino acid. At least to a first approximation, the peptide interactions, solvent effects, inside/outside distribution of amino acids, presence of disulfide bonds and all other such complexities are neglected. This approach is commonly used to estimate the macroscopic phase transition properties of complex hydrocarbon mixtures (e.g., the enthalpy of vaporization of petroleum distillates) from the properties of the components of the mixture. This approach also lends itself to modeling the effects of hydrophobic interactions on protein stability because the hydrophobic interaction is not a true chemical bond with all the inherent steric constraints of such bonds.

If we wish to predict protein stability using a phase transition model, then we must in turn predict the thermodynamic variables upon which stability depends. For the simplest case, that of a reversible two-state unfolding transition, the

Gibbs-Helmholtz equation (7) relates the enthalpy and heat capacity change upon unfolding, Δh_u and Δc_p^u, respectively, to protein stability (the free energy change of protein unfolding, $\Delta g(T)$) as follows:

$$\Delta g(T) = \Delta h_u \left(\frac{T_u - T}{T_u}\right) - \int_T^{Tu} \Delta c_p^u \, dT + T \int_T^{Tu} \Delta c_p^u d(\ell nT) \qquad (1)$$

In this equation $\Delta g(T)$ is the specific free energy change of unfolding, J (gm. protein)$^{-1}$, under given conditions, T_u is the temperature (K) at which unfolding occurs under these conditions and T is the system temperature (K). Thus the minimum information required to rigorously calculate $\Delta g(T)$, the protein stability, is T_u, Δh_u and Δc_p^u. The unfolding temperature is frequently taken as the primary measure of protein stability; in fact, it is only one of three parameters needed to rigorously calculate protein stability for the simplest case of protein unfolding. While the unfolding temperature is relatively easy to measure; Δh_u and Δc_p^u are not. If Δh_u and Δc_p^u can be predicted, then they can in turn be used to predict the protein stability based on Equation 1 or similar relationships. The model developed here predicts Δh_u and Δc_p^u, not T_u.

Predicting the Enthalpy and Heat Capacity Change of Protein Unfolding. The definition of the hydrophobic interaction given by Dill (6) is used, namely the transfer of a nonpolar solute into aqueous solution with an accompanying large positive change in the heat capacity. One operational difficulty with this definition is the choice of the solvent from which the transfer of the nonpolar solute to aqueous solution is made. The approach taken here was to correlate many sets of amino acid properties, including properties related to amino acid transfer from various solvents to water, with experimental values of Δh_u and Δc_p^u. The choice of the "best" solvent was then purely pragmatic. The best solvent was the one which gave the best prediction of the experimental values of Δh_u and Δc_p^u. The literature was searched to find sets of amino acid property data and to locate enthalpy and heat capacity change data for compact globular proteins of a single domain, since compact, single domain proteins are required for the simple two state phase equilibrium model chosen for study here. Furthermore, the calorimetric data itself should demonstrate that the protein has a single reversibly unfolding domain with approximately constant heat capacity change and the amino acids comprising the domain must be known. The experimental values of enthalpy and heat capacity change for all ten proteins meeting these criteria are summarized in Table 1 together with one of the normalized phase properties (described below) which we call the transfer free energy density.

Therefore, a simple algorithm was used to calculate a weighted (by amino acid content) average value of the various amino acid properties for all ten proteins for which enthalpy and heat capacity change data were available and which also exhibit reversible, two-state unfolding. The algorithm was

$$P_{i,j} = \sum n_{k,j} \cdot P_{i,k} \qquad (2)$$

where $P_{i,j}$ is the weighted value of property i in protein j, $n_{k,j}$ is the number of the kth amino acid residues in the jth protein and $P_{i,k}$ is the value of the ith property for the kth amino acid. The model (algorithm) thus makes no provision for primary structure, for disulfide bond content or for any other protein property, only overall amino acid composition is used. In the language of thermodynamics, this model

TABLE 1

EXPERIMENTAL ENTHALPIES AND HEAT CAPACITIES OF PROTEIN
UNFOLDING AND THE
CALCULATED TRANSFER FREE ENERGY DENSITY

PROTEIN (Abbreviation)*	ENTHALPY CHANGE (J/g at 25°C) Experimental	HEAT CAPACITY CHANGE (J/g · K) Experimental	TRANSFER FREE ENERGY DENSITY (J/g) Calculated	EXPERIMENTAL DATA FROM REFERENCES
Bovine pancreatic ribonuclease (Rna)	22.1	0.38	26.8	*3,8*
Hen egg white lysozyme (Lys)	13.2	0.46	30.6	*8,9*
Sperm whale metmyoglobin (Myo)	2.30	0.64	39.1	*3,8,10*
Bovine β trypsin (Trp)	10.9	0.50	34.9	*3*
Bovine cytochrome c (Cyt)	5.90	0.61	36.7	*3,8*
Bovine cytochrome b_5 (frag. 1-90) (B5)	7.45	0.55	31.1	*11,12*
Rabbit liver carbonic anhydrase (Car)	5.06	0.58	38.6	*3*
Bovine α chymotrypsin (Chy)	11.1	0.50	35.4	*8,13*
Carp muscle parvalbumin (Par) (without Ca^{++})	9.79	0.49	36.4	*14*
Chicken erythrocyte histone H5 (GH5) (frag. 22-100)	2.89	0.58	35.1	*15*

* These abbreviations were used to designate the proteins in Fig. 1.

treats the protein molecule as a phase composed of an ideal solution of amino acids. That is, the overall property of the phase is a linear additive function of the constituents of the phase, i.e., of the amino acids.

The available experimental data are usually specific enthalpy changes rather than molar enthalpy changes. Thus the resulting value of $P_{i,j}$ was averaged by dividing by protein molecular weight. These normalized phase property values, now denoted $\overline{P}_{i,j}$, were then tested for possible correlation versus experimental values of Δh_u (at 25°C) and Δc_p^u by linear curve-fitting. Table 2 summarizes the results of a linear fit of the various $\overline{P}_{i,j}$ versus Δh_u (at 25°C) and Δc_p^u for all ten proteins for those properties which achieve correlation coefficients greater than 50%. Only for correlation coefficients greater than 50% can the hypothesis that there is no correlation between the $\overline{P}_{i,j}$ and Δc_p^u or Δh_u be rejected with greater than 90% confidence. Properties which achieved correlation coefficients of less than 50% were: transfer free energy from N-methylacetamide to water and cyclohexane to water (18,24), hydrophilicity (25), various hydrophobicity parameters (20,23,26,27), hydropathy (28), measures of fractional burying and accessibility (20,29,30), free energy of side chain hydration (31), volume, contacts per residue and turn potential (27), bulkiness (32) and normalized partition energies from n-octanol to water (33) .

The Transfer Free Energy Density Model. Of the various solvents tested (95% ethanol in water, anhydrous ethanol, n-octanol, cyclohexane, N-methylacetamide and dioxane), 95% ethanol in water achieved by far the best prediction of experimental Δh_u and Δc_p^u values. We refer to this empirical relation (or correlation) between the computed ethanol to water transfer free energy per gram of protein and the experimental values of enthalpy and heat capacity change as the transfer free energy density (TFE) model.

Since the heat capacity change on protein unfolding for compact, globular proteins is known to be directly proportional to the concentration of nonpolar contacts (3), it is reasonable to postulate that the protein heat capacity change upon unfolding will be zero if the hydrophobic contact density is zero. In terms of the transfer free energy density model, Δc_p^u will be zero when the transfer free energy density is zero. (This postulate also requires that Δh_u (25°C) be 54.4 J/g at zero transfer free energy density.) When these two values are included in the correlation, the correlation coefficients (R values) for Δc_p^u and Δh_u versus TFE density are 0.97 and 0.98, respectively. The experimental calorimetric data for all ten proteins are plotted in Figure 1 versus the computed TFE density. The lines drawn through the data represent the best linear fit of the data including the postulated zero points for Δc_p^u and Δh_u (25°C), i.e., 11 data pairs in total for both heat capacity and enthalpy change. (The trend of the experimental data also clearly supports the postulate that the heat capacity change will be zero at zero transfer free energy density.)

The probability that the linear correlation observed here between the computed TFE values and the experimental thermodynamic parameters is not significant is less than one in one million (34), given a correlation coefficient of 0.97. If the postulated zero points are not included in the correlation, the correlation coefficient is approximately 0.80 (Table 2). Given this value of the correlation coefficient, the probability that this is not a significant correlation is about one in one thousand. Thus these empirical correlations must be considered meaningful, regardless of whatever fundamental mechanism(s) one might wish to propose as the basis for their empirical success.

TABLE 2

CORRELATION COEFFICIENTS (R VALUES AS
PERCENT) FOR LINEAR FIT OF CALORIMETRIC DATA
AND ALGORITHM RESULTS $(\overline{P}_{i,j})$

AVERAGE PROPERTY	VERSUS ENTHALPY CHANGE	VERSUS HEAT CAPACITY CHANGE	PROPERTY DATA FROM REFERENCE
Transfer free energy (95% ethanol to water at 25°C)	83	80	16
Solubility in water at 25°C	73	68	17
Transfer free energy (ethanol to water at 37°C))	70	72	18
Transfer free energy (ethanol to water at 25°C)	70	69	19
Hydrophobic index	68	69	20
Water molecule ordering	57	54	21
Transfer free energy (to surface at 30°C)	59	51	22
Logarithm of distribution coefficient (octanol/water)	50	58	23
Polarity	45	52	20

The equations for predicting Δh_u and Δc_p^u (or in other words, the equations of the straight lines in Figure 1) are:

$$\Delta h_u \ (25°C) = 54.4 - 5.48 \times (TFE) \qquad (J/g) \qquad\qquad (3)$$

$$\Delta c_p^u = -0.00322 + 0.0644 \times (TFE) \qquad (J/g·K) \qquad\qquad (4)$$

where (TFE) is calculated from Equation 2 using the data of Stellwagen (16) for $P_{i,k}$ (95% ethanol to water transfer free energies, values given below) and the amino acid composition of the protein of interest, $n_{k,j}$. The resulting quantity is divided by the protein molecular weight and converted to units of J/g protein to obtain (TFE) which is then substituted in Equations 3 and 4 to obtain Δh_u and Δc_p^u.

A Test of the TFE Model.   One readily accessible test of the model is to calculate the enthalpy change of unfolding at the temperature of unfolding (assuming constant Δc_p^u) for the ten proteins in the data base using Equations 3 and 4 and then to compare the predicted (calculated) values with the experimental results. This comparison between the model predictions and experiment was performed and the results are summarized in Table 3. The average absolute difference between the calculated and experimental values is 9.2%, which is probably within experimental error.

Preliminary Conclusions Derived from the TFE Density Model.   Several preliminary but nonetheless important conclusions result from this successful prediction of the macroscopic properties of protein unfolding from a computed parameter based only on the amino acid composition. First, success of this correlation is consistent with the idea of protein unfolding as a process dominated by transfer of amino acids from a hydrophobic environment (the solvent) to water, as has often been postulated before (see for instance, 2,6).

Second, it is surprising to note that although five of these proteins (bovine cytochrome b_5, (fragment 1-90), sperm whale metmyoglobin, chicken erythrocyte histone H5, (fragment 22-100), carp muscle parvalbumin and bovine carbonic anhydrase) lack disulfide bonds, their experimental values of Δh_u and Δc_p^u correlate with the computed TFE in the same way as proteins which have one or more disulfide bridges. The presence (or absence) of disulfide bonds does not appear to affect Δc_p^u or Δh_u if these bonds remain intact during the unfolding process. Apparently the influence of disulfide bonds, if any, on stability ($\Delta g(T)$) must be manifested primarily through T_u. This is also consistent with the notion that the primary influence of disulfide bonds on protein stability is to increase stability by decreasing the entropy of the unfolded protein. Data for T4 lysozyme mutants tend to support the idea that the presence of or absence of disulfide bridges primarily affects the unfolding temperature (35,36).

Third, there has been considerable discussion about which are the "hydrophobic" amino acids. Of the twenty common amino acids, only eleven in this model have non-zero transfer free energies, i.e., only eleven are hydrophobic. The side chain transfer free energy values used in the TFE model are those summarized or assigned by Stellwagen (16), namely: Ala, Arg (0.6 kcal/mol), Lys, Met, Val (1.5 kcal/mol) and Ile, Leu, Phe, Pro, Trp, Tyr (2.5 kcal/mol).) All other amino acid side chains were assigned 95% ethanol in water to water transfer free energies of zero, i.e., the other nine amino acids are not "hydrophobic". Although it seems counterintuitive, it is nonetheless true that this phase equilibrium approach using these transfer free energy values provides qualitatively better predictions of Δh_u and

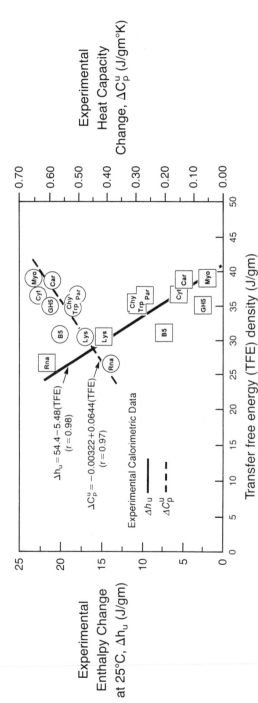

Figure 1. Experimental thermodynamic data versus computed transfer free energy density values for ten proteins.

TABLE 3

EXPERIMENTAL VERSUS PREDICTED

VALUES OF ΔH AT THE UNFOLDING TEMPERATURE

PROTEIN	UNFOLDING TEMPERATURE °C	EXPERIMENTAL ENTHALPY CHANGE J/g	PREDICTED ENTHALPY CHANGE J/g
Bovine pancreatic ribonuclease	64	36.9	35.5
Hen egg white lysozyme	77	39.1	39.0
Sperm whale metmyoglobin	78.5	36.4	35.6
Bovine β trypsin	25*	45.6	36.3
Bovine cytochrome c	78	38.2	36.4
Bovine cytochrome b_5	70.7	32.6	35.6
Rabbit liver carbonic anhydrase	60	28.2	24.8
Bovine α chymotrypsin	57	27.1	25.5
Carp muscle parvalbumin	35	14.7	12.2
Chicken erythrocyte histone H5	63	25.0	29.0

* extrapolated value

Δc_p^u than does an identical algorithm in which all twenty side chains (except for glycine) have non-zero transfer free energies from n-octanol to water. Therefore, strictly in terms of its empirical ability to predict these macroscopic thermodynamic properties, 95% ethanol in water is obviously superior to n-octanol in representing the protein "phase".

Fourth, protein fragments and complete proteins obey the same correlation, emphasizing that Δc_p^u and Δh_u are determined by a compact globular core (*15*). If we take obedience to this correlation as the criterion defining an independently folding unit of the protein, i.e., as a defining a protein domain, then we can attempt to predict the unfolding behavior of complex, multidomain proteins by treating them as ensembles of individual domains. The remainder of this paper deals with the use of Equations 3 and 4, the TFE model, to predict protein stability. Special attention is given to protein stability in high temperature environments. Recall that the TFE model only predicts Δh_u and Δc_p^u, the unfolding temperature must be experimentally determined or estimated using some other model.

Predictions of Protein Stability using the TFE Model

Introduction. As discussed above, the minimum information required to rigorously calculate protein stability is the enthalpy and heat capacity change of protein unfolding and the unfolding temperature, T_u. The enthalpy and heat capacity change are predicted from Equations 3 and 4, given the composition of the domain. The unfolding temperature, if not available from experiment, is estimated using the model of Bull and Breese (*37*). The protein stability as a function of temperature is then calculated from Equation 1, assuming the heat capacity change to be approximately constant with temperature, and the resulting predicted stabilities are tested with experimental information from the literature for various proteins.

Lambda Phage Repressor Protein. This protein consists of two independent domains, an amino terminal fragment (1-92) and a carboxy terminal fragment (132-236). The following comments refer to Figure 2. In Figure 2, the C domain is predicted by the TFE model to be 2.9 kcal/mol more stable than the N domain at 292 K. Calorimetric information shows that the N terminal domain unfolds at about 50 C while the C terminal domain unfolds at about 70 C (*38*). Likewise in Figure 2 the G48N mutant is predicted to be approximately 0.7 kcal/mol more stable than the wild type at 53 C while Hecht, et al, (*39*) experimentally determine the mutant to be approximately 0.6 kcal/mol more stable than the wild type at this temperature. Finally, the E34K mutant is predicted by the model to be slightly less stable than the wild type and this is borne out by the observation of the experimentalists that there is a "small destabilizing effect of this substitution" (*39*).

The stabilities of the amino terminal domain for the wild type repressor protein and several other mutant proteins have also been experimentally measured (*40*). These experimental values of the free energy of stabilization are compared in Table 4 with the values predicted by the TFE model and the experimentally determined unfolding temperatures. As this table shows, the agreement between the model predictions and experiment is excellent. Unfortunately, near the unfolding temperature, the temperature of unfolding completely dominates the computed stability, and hence the heat capacity and enthalpy change values predicted by the model are not well-tested in this region. For this reason the extensive data base on stability of T_4 lysozyme mutants (*35,36*) cannot be used to test the predictions of the TFE model. All the change in stability data, $\Delta(\Delta g(T))$, for these T_4 lysozyme mutants were evaluated very near the unfolding temperature.

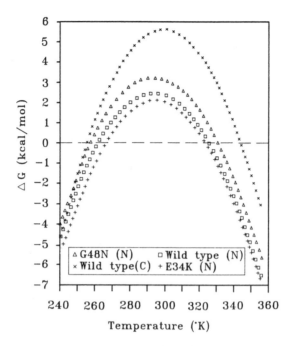

Figure 2. Predicted stabilities of the N and C domains of lambda phage repressor protein and two mutants.

TABLE 4

**CHANGE IN GIBBS FREE ENERGY OF UNFOLDING OF
THE AMINO TERMINAL DOMAIN OF PHASE λ REPRESSOR
AND ITS MUTANT PROTEINS**

PROTEIN	Td (°C)	ΔG(51.5°C) TFE Model	(kcal/mol) Measured	% DIFF
Wild	51.5	0	0	-
Q33Y	57.4	0.90	1.32	32
A49V	38.5	-1.58	-1.22	-30
Y22H	28.8	-2.53	-2.52	-12
A66T	29.0	-2.42	-2.99	19
I84S	37.2	-1.83	-2.25	19

The available calorimetric data on tryptophan synthetase alpha subunit, staphylococcal nuclease and their mutants can also not be used to test the TFE model because one or more of the conditions required for use of the model are not met. However, as one moves away from this region near the unfolding temperature, the other thermodynamic parameters of enthalpy and heat capacity change also significantly affect the computed stability.

Lactate Dehydrogenase. Lactate dehydrogenase is also a two domain protein with known domain limits (*41*). The estimated stabilities of these two domains based on the TFE model for the enzymes derived from pig and dogfish are displayed in Figure 3. These predicted stabilities correspond well to the observation in the literature that the pig M4 N domain is more stable than the dogfish M4 N domain (*42*).

Ferridoxins from Mesophiles and Thermophiles. Amino acid composition information is available on ferridoxins from both thermophiles and mesophiles (*42*). These compositions were used to predict the enthalpy and heat capacity change for protein unfolding from Eqs. 3 and 4 above. The Bull and Breese model was used to predict the temperature of unfolding for each of the proteins. The computed $\Delta g(T)$ values are displayed in Figure 4. It is important to note that while the Bull and Breese model does generally predict an increased unfolding temperature for the thermophiles, the predicted increases in T_u are quite small. Nonetheless, a clear pattern emerges that the $\Delta g(T)$ curves for ferridoxins from thermophiles are significantly higher than those from mesophiles over the physiologically relevant temperature ranges, i.e., the thermophilic proteins are indeed predicted by the model to be more stable than the mesophilic proteins.

Proteins from Hyperthermophiles and the TFE Model. Much of this volume is concerned with hyperthermophiles and proteins from these organisms. Very little thermodynamic information on these proteins is available as yet. However, the TFE model in conjunction with the Bull and Breese model can be used to determine patterns of amino acid replacement and other protein properties which would be predicted to generate hyperthermophilic behavior. These predictions can then be compared with experiment as such experimental data continue to accumulate. Kanamycin nucleotidyltransferase (KNTase) was used as the model protein for these simulations since the relative stability of mutants of this protein at position 80 correlate well with the hydrophobicity of the substituted residues as measured by ethanol to water transfer free energies (*43*).

Figure 5 shows the hypothetical effect of replacing all six asparagine residues in the first 80 residues of KNTase with six leucines. In other words, this simulation greatly increases protein hydrophobicity but has little effect on molecular weight. The temperature at which Δg equals zero is shifted up by this substitution, i. e., the protein becomes more thermostable. The temperature of maximum stability (maximum positive free energy change) is also shifted up by this hypothetical "mutation" as is the temperature of cold denaturation (*44,45*). Therefore the model predicts that the protein becomes more cold labile at the same time it is becoming more thermostable.

Figure 6 shows the effect of doubling the molecular weight of KNTase (1-80) for the "mutant" with six leucines while holding constant the amino acid composition. The effect of this simulation is to hold the hydrophobicity of the protein constant while doubling its molecular weight. These two curves for the larger and smaller protein are symmetrical and intersect the Δg equals zero axis at the same high and low temperatures. However, there are two major effects of doubling

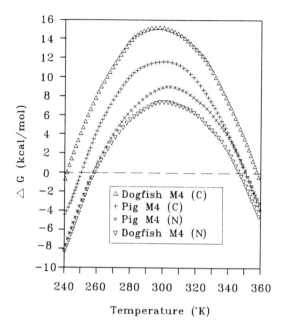

Figure 3. Predicted stabilities of the N and C domains of two lactate dehydrogenases.

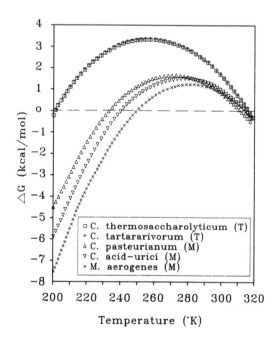

Figure 4. Predicted stabilities of ferridoxins from thermophiles and mesophiles.

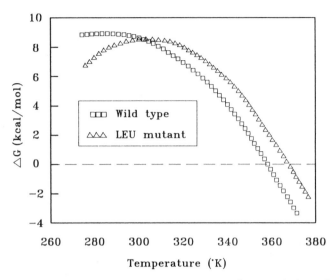

Figure 5. Predicted stabilities of hypothetical domains consisting of the first 80 residues of kanamycin nucleotidyl transferase and a leucine-rich mutant.

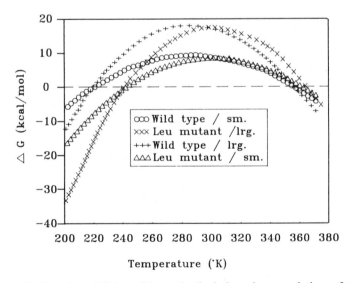

Figure 6. Predicted stabilities of hypothetical domains consisting of the first 80 residues of kanamycin nucleotidyl transferase and a leucine-rich mutant: effect of doubling domain size.

the molecular weight: 1) the maximum stability of the protein is doubled along with its molecular weight (an obvious but nonetheless important result) and 2) the stability of the larger protein increases more rapidly with decreasing temperature near the high temperature unfolding transition than does the stability of the smaller protein, i.e., even at high temperatures the larger protein rapidly acquires significant stability as the temperature decreases somewhat from the unfolding temperature.

Prediction of the Properties of Proteins from Hyperthermophiles. Based on these simulation results and the results from Figure 1, hyperthermophilic proteins are predicted to have the following properties: 1) they will <u>not necessarily</u> have many or even any disulfides, 2) they may have larger domains on average than proteins from mesophiles, 3) they will probably have an increased content of hydrophobic amino acids (the hydrophobic amino acids are taken to be the eleven amino acids with non-zero ethanol to water transfer free energies), 4) they will tend to be cold labile and prone to aggregation, and 5) otherwise, these proteins will not be greatly different than their counterparts from mesophilic organisms.

These predictions were made prior to reports from Dr. Bob Kelly's group (see article in this volume) which confirm that the alpha glucosidase from *Pyrococcus furiosus* does indeed have many of the properties that the TFE model predicts. This enzyme: 1) has no disulfides, 2) is apparently a single domain with a molecular weight of approximately 150,000 Daltons (proteins from mesophiles tend to have domains in the range of 10-20 kD), 3) it has a computed heat capacity change of unfolding of approximately 0.60 J/g/K, near that of sperm whale metmyoglobin, which is the most hydrophobic of the ten proteins in the data base for the TFE model (see Table 1), and 4) this alpha glucosidase is very prone to aggregation. It is also predicted to be extremely stable; the TFE model estimates a maximum free energy of stability at 25 C of approximately 110 kcal/mol, five to ten times as high as the predicted maximum stability of most of the other proteins discussed here. The large size of the protein is responsible for most of this extreme stability. The Bull and Breese model only predicts an unfolding temperature of about 81 C, which is not consistent with the extreme thermostability of this enzyme.

Conclusions

A phase equilibrium approach to protein stability has been used to generate correlations (the TFE model) which well predict the enthalpy and heat capacity change upon unfolding for compact globular proteins of a single domain. These thermodynamic parameters, used in conjuction with experimentally-determined unfolding temperatures or unfolding temperatures estimated from the Bull and Breese model, can in turn be used to predict protein stability as a function of temperature. Preliminary tests of these computed stabilities versus experimental protein stabilities have proven reasonably successful. The model also predicts some of the emerging features of proteins from hyperthermophiles. The Bull and Breese model seems to work reasonably well for moderate temperatures of unfolding, it does not well predict the high unfolding temperature for the alpha glucosidase of *Pyrococcus furiosus*.

No molecular level interpretations are claimed (or indeed are even possible) based on this model. As with all other models, the TFE model will be found to be useful in some cases and of little or no value in others. Nonetheless, this first generation model does provide a simple means of estimating important thermodynamic information related to protein stability. It may be possible to extend the model to account for pH effects by using appropriate pH dependent transfer free energy values.

Acknowledgments

B. Dale is grateful to the National Institute of Standards and Technology for support as a visiting scientist during his sabbatical year, during which much of this work was accomplished, and to the National Science Foundation for partial support (CBT-8815543).

Literature Cited

1. C. Ghelis and J. Yon., Protein Folding, p. 180, Academic Press (1982).
2. Kauzmann, W., *Adv. Protein Chem.*, **14**, 1, (1959).
3. P. L. Privalov, *Adv. Protein Chem.*, **33**, 167 (1979).
4. R. L. Baldwin, *Proc. Nat. Acad. Sci. USA*, **83**, 8069 (1986).
5. K. Murphy, P. L. Privalov and S. J. Gill, *Science*, **247**, 559 (1990).
6. K. A. Dill, *Biochemistry*, **29**, 7133 (1990).
7. W. Pfeil, *Mol. and Cell. Biochem.*, **40**, 3 (1981).
8. P. L. Privalov and N. N. Khechinashvili, *J. Mol. Biol.*, **86**, 665 (1974).
9. N. N. Khechinashvili, P. L. Privalov and E. I. Tiktopulo, *FEBS Letters*, **30**, 57 (1973).
10. P. L. Privalov, N. N. Khechinashvili and B. P. Atanasov, *Biopolymers*, **10**, 1865 (1971).
11. P. Bendzko and W. Pfeil, *Biochim. Biophys. Acta*, **742**, 669 (1983).
12. W. Pfeil and P. Bendzko, *Biochim. Biophys. Acta*, **626**, 73 (1980).
13. V. M. Tischenko, E. I. Tiktopulo and P. L. Privalov, Biofizika, **19**, 400 (1974).
14. V. V. Filimonov, et al, *Biophys. Chem.*, **8**, 117 (1978).
15. E. I. Tiktopulo, et. al., *Eur. J. Biochem.* **122**, 327 (1982).
16. E. Stellwagen, *Ann. N.Y. Acad. Sci.*, **7**, 1 (1984).
17. G. D. Fasman (ed.), Handbook of Biochemistry and Molecular Biology, 3rd Ed., Vol. II, p. 188, CRC Press, Boca Raton, Fl. (1984).
18. S. Damodaran and K. B. Song, *J. Biol. Chem.*, **261**, 7220 (1986).
19. Y. Nozaki and C. Tanford, *J. Biol. Chem.* 246, (7), 2211 (1971).
20. P. K. Ponnuswamy, M. Prabhakaran and P. Manavalan, **623**, 301 (1980).
21. G. Nemethy and H. A. Scheraga, *J. Chem. Phys.*, **66**, 1773 (1962).
22. H. B. Bull and K. Breese, *Arch. Biochem. Biophys.*, **161**, 665 (1974).
23. J. L. Fauchere and V. Pliska, *Eur. J. Med. Chem. -Chim. Ther.*, **18**, 369 (1983).
24. A. Radzicka and R. Wolfenden, *Biochem.*, **27**, 1664 (1988).
25. T. P. Hopp and K. R. Woods, *Proc. Nat. Acad. Sci. USA*, **78**, 3824 (1981).
26. M. Levitt, *J. Mol. Biol.*, **26**, 59 (1976).
27. R. Ragone, et al, *Ital. J. of Biochem.*, **36**, A306 (1987).
28. J. Kyte and R. F. Doolittle, *J. Mol. Biol.*, **157**, 105 (1982).
29. C. Chothia, *J. Mol. Biol.*, **105**, 1 (1976).
30. G. D. Rose, et al, *Science*, **229**, 834 (1985).
31. B. Robson and D. J. Osguthorpe, *J. Mol. Biol.*, **132**, 19 (1979).
32. D. D. Jones, *J. Theor. Biol.*, **50**, 167 (1975).
33. H. R. Guy, *Biophys. J.*, **47**, 61 (1985).
34. G. W. Snedecor and W. G. Lochron, Statistical Methods, p. 477, Iowa State Univ. Press. (1980).
35. T. Alber et al., Nature, **330**, 41 (1987).
36. B. W. Matthews, H. Nicholson, and W. J. Becktel, *Proc. Natl. Acad. Sci., USA*, **84**, 6663 (1987).
37. H. B. Bull and K. Breese, *Arch. Biochem. Biophys.*, **158**, 681-686, (1973).

38. Pabo, *Proc. Natl. Acad. Sci. (USA)*, **76**, 610 (1979).
39. Hecht, M. H., Hehir, K. M., Nelson, C. M. Sturtevant, J. M. and Sauer, R. T., *J. Cellular-Biochem.*, **29**, 217 (1985).
40. Hecht, M. H., Sturtevant, J. M., and R. T. Sauer, *Proc. Natl. Acad. Sci. (USA)*, **81**, 5685 (1984).
41. Kiltz, H. H., Keil, W., Griesbach, M., Petry, K. and Meyer, H., *Hoppe-Seyler's 2. Physiol. Chem.*, **358**, 123 (1977).
42. Argos, P., Rossman, M. G., Grau, U. M., Zuber, H., Frank, G., and Tratschin, J. D., *Biochem.*, **18**, 5698 (1979).
43. Matshumura, M., Yahanda, S., Yasumura, S., Yutani, K. and Aiba, S., Eur. *J. Biochem.* **177**, 715 (1988).
44. Griko, Y. V., Privalov, P. L., Sturtevant, J. M., and Venyaminov, S. Y., *Proc. Natl. Acad. Sci (USA)*, **85**, 3343 (1988).
45. Privalov, P. L., Griko, Y. V. and Venyaminov, S. Y., *J. Mol. Biol.*, **190**, 487 (1986).

RECEIVED January 24, 1992

Chapter 11

Computational Approaches to Modeling and Analyzing Thermostability in Proteins

John E. Wampler[1], Elizabeth A. Bradley[1], Michael W. W. Adams[1], and David E. Stewart[2]

[1]Department of Biochemistry and [2]University Computer and Network Services, University of Georgia, Athens, GA 30602

Both experimental and theoretical studies are beginning to reveal factors in protein structure which lead to thermal stability. This paper focuses on computational approaches such as analysis of sequence and structure databases, molecular modeling of thermostable proteins using known X-ray structures, molecular mechanics and dynamics calculations, and free energy perturbation calculations. Analysis of database information has not yet revealed a clear composition or sequence basis for thermostability in proteins. However, homology modeling and molecular dynamics simulations do appear to be having some success in probing the effects of mutations on protein thermostability. The results of free energy calculations also indicate that predictions can be made that correlate well with experimental data.

The problem of protein thermostability is in reality a number of separate and distinct problems as illustrated in Figure 1. Even the reversible processes (Type I) can be complex involving dissociation of subunits and multiple steps of unfolding. The simplest reversible process, seen with small, single chain globular proteins, can be modeled as a two state phase change (*1*). With larger proteins or proteins involving multiple subunits, reversible thermal denaturation is composed of multiple steps (*1-4*). Details of the steps involved can come from a variety of experimental studies (see below), but, a good overall indicator of the Type I events can be measured in the form of the melting temperature, Tm. In the simple case, Tm is the temperature where half of the protein is in the folded state and half in the unfolded state. In more complex systems, the Tm is the inflection point of the melting curve or the point where half of the total change has occurred.

Melting temperatures of proteins are highly dependent on environmental factors, pH, solute concentrations, etc. The Tm for lysozyme at pH 2 is 42°C,

0097–6156/92/0498–0153$06.25/0

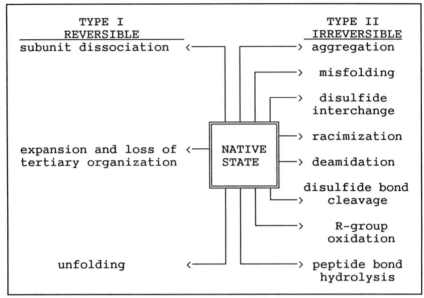

Figure 1. Mechanisms of Thermal Denaturation for Proteins.

but increases to 65°C at pH 6.5 (5). Yeast alcohol dehydrogenase exhibits a near linear dependence of Tm with pH from a low of 45°C at pH 9.4 to a high of 74°C at pH 6.2 (6) and its stability is modulated by the presence of its cofactor, NAD^+. With other electron transfer proteins containing prosthetic groups, the oxidation state can make a considerable difference in stability. The Tm of cytochrome b_{562} from Escherichia coli is 81°C in the reduced state and 67°C in the oxidized state (7).

In contrast to the reversible process which is typically intramolecular and driven by fast kinetic events, the slower, irreversible processes (Type II in Figure 1) are primarily chemical (see references 8 and 9). The Type II events are typically measured at the kinetic half-life, $t_{\frac{1}{2}}$, at high temperature. Major contributions to this process have been identified as peptide bond hydrolysis at aspartyl residues, β-elimination of cystine and deamidation of asparagine. Amide hydrolysis at aspartyl residues is accelerated at acid pH's while the β-elimination reaction is catalyzed by base. Other sources of irreversible denaturation are: oxidation of amino acid side chains (particularly methionine and cysteine), racimization of amino acids, and disulfide interchanged catalyzed by thiols whether from the ß-elimination of cystine or some exogenous source. In fact, racimization of a number of amino acid residues occurs at rates that are faster within proteins than for the free amino acid.

Much of the work on extremely thermophilic proteins reports their stability in terms of the half-life of the protein when incubated at a high temperature as determined by the loss of catalytic activity (typically at 85°C, $t_{\frac{1}{2}}^{85}$, or at 95°C, $t_{\frac{1}{2}}^{95}$). These measurements in many cases, perhaps most, reflect a combination of both Type I and Type II denaturation events. The tests for purely Type I-- reversibility and kinetics which indicate an approach to equilibrium rather than first order decay-- are typically not reported in these studies. Incubation times at high temperatures in a typical experiment are often long enough for significant chemical change (*10*). Table I gives some half-life values for proteins from extreme thermophiles. For comparison, the half-life for deamidation of asparagine residues at 100°C (pH 6) is 0.169 hours for lysozyme (*10*) and about 0.25 hours for triosphosphate isomerase under the same conditions (*11*). One obvious solution to this particular problem would be for proteins from extreme thermophiles to have lower asparagine content. However, there is no evidence to date that supports this or any other consistent difference in composition between proteins from extreme thermophiles and mesophiles (see below).

Experimental Measurements

Pace et al. (*12*) have summarized some of the experimental approaches to measuring conformational stability of proteins and analyzing the data. Experimentally, it can be difficult to distinguish between the reversible and irreversible events, and it is reasonable to assume that many measurements reflect the combination of both effects. The best test of reversibility is reversibility!

In kinetic measurements of the decay of protein activity or of changes in spectroscopic properties at high temperatures, a single term first order process represents an irreversible reaction or reversible denaturation at a temperature well above the Tm where the new equilibrium involves nearly complete conversion. Decay to a new equilibrium, non-zero position might, on-the-other-hand, represent a reversible process. Multi-component decays can arise from a mix of multiple Type I and/or Type II processes.

The measurements which can most directly be correlated with computational approaches are thermodynamic measurements from differential scanning microcalorimetry (*1*) and analysis of melting curves (*12*). Microcalorimetry gives a detailed view of the transistions involved in denaturation and of the properties of each discrete state.

It should be noted that the literature has many reports of "Tm" and "$t_{\frac{1}{2}}$" values that are not from scanning microcalorimetry and kinetic decay measurements. Bull and Breese (*22*) and Stellwagen and Wilgus (*23*) report melting temperatures, "Tm's", that are obtained by examining pH change with temperature. However, these experiments typically show irreversible denaturation in the formation of protein aggregates at the transition temperatures. Stellwagen and Wilgus (*24*) suggest that these temperatures "should be considered as a scanning analog of kinetic measurements." Similarly, the half-life values used by Guruprasad et al. (*25*) are *in vivo* half-life values and contain information on protein synthesis and enzymatic degradation as well as stability.

Table I. Half-Lifes for Some Proteins from Extreme Thermophiles   *

Temperature (°C)

Protein	80	85	90	95	100	reference
α-Amylase	21			>1		13
L-Asparaginase		0.67		0.04		13
D-Asparaginase		0.42				13
Caldolysin		>5		>1		13
Carboxymethyl cellulase		1				13
β-Galactosidase		>5		0.03		13
α-Glucosidase					~47	14
β-Glucosidase		0.02				13
Glyceraldehyde Phosphate Dehydrogenase					0.73	15
Ferredoxin				>24		16
Hydrogenase	21				2	17
Isocitrate dehydrogenase			0.55		0.03	18
Malate dehydrogenase					0.06	18
NADH dehydrogenase		>5		0.67		13
Phospho-glycerate kinase			0.17			19
Protease (Crude)					33**	20
Rubredoxin				>24		21

*Values in hours.   ** 98°C rather than 100°C

Many workers (eg. *26, 27*) have assumed that the melting temperatures of proteins will reflect the natural or optimal growth temperatures of the organisms from which the proteins were obtained.   However, several examples of proteins from mesophilic organisms have been found that exhibit a high degree of thermostability.   For example, the tailspike protein from bacteriophage P22 has a melting temperature of 88°C (*28*) and the rubredoxin from the mesophile, <u>Desulfovibrio</u> <u>gigas</u>, has a spectroscopic half-life of 2 hours at 80°C (*29*).

Thermodynamic Considerations

One question that challenges the possibility of learning anything useful about thermal stability of proteins from computational approaches is the fundamental question of the thermodynamics involved.   Reversible denaturation events at moderate temperatures have been viewed for some time as entropy driven and the stability of native conformations has been explained by the entropy gained from elimination of solvent as the hydrophobic core folds (see *30*).   However, in addition to this entropy gain, there is also a large entropy loss due to the many conformational states that the unfolded protein chain can assume.   So there are two competing effects whose entropy contributions to the total free energy are on the order of hundreds of thousands of calories per mole according to various measures and calculations (see *30, 31*).   In addition, there are obviously a number of enthalpic components involved in protein folding.   The totals of both the positive and negative enthalpic contributions are also large. In contrast to all of these large energy terms, the typical net free energy for the unfolding transistion of a small, globular protein averages approximately 10 Kcal/mole (*32*).   Thus, the uncertainty in the exact magnitude of these large energy contributions has allowed a fundamental argument about the energetics of protein stability to remain unresolved.   Current analysis using microcalorimetry data (*1*) suggests that the temperature for reversible denaturation is controlled by intramolecular interactions which stabilize the folded form of a protein, especially since solvation and entropic effects on the unfolded form are not very sensitive to composition.   Therefore, analysis of the enthalpies of interactions within a protein may well provide pertinent information for understanding its thermostability.

Thus, the free energy difference between the folded and unfolded states of soluble globular proteins is on the order of the energy contribution for a few hydrogen bonds or ionic interactions (see *33* for a review) and it represents a small difference between a number of large energy terms.   This makes theoretical calculations of the net stabilizing energy difficult because of the precision required.   For example, calculations of the entropy of the folded state can differ by a factor of two depending on the approach (see *31*).   The total contribution of this term to the free energy is large.   For a small protein at room temperature it can be several hundred Kcal/mol (see *31*, pp. 180 ff).   Thus, the uncertainty in the calculation is greater than the net free energy that is sought.   For these reasons, a purely theoretical approach to the question of protein stability using energy calculations of the component contributions on an absolute scale is unlikely to yield definitive answers. The question is then typically narrowed to one of differences in stability between closely analogous proteins, mutants or natural variants, using experimental and theoretical procedures correlated on a relative, rather than absolute scale.

Evidence in support of the idea that evaluations of interaction energies might be useful to studying thermal stability comes from a number of examples where single, small changes in a protein can be seen to have very significant impact on their stability. For example, single site mutations have, in several cases, been demonstrated to cause a 1-2 Kcal/mole change in the free energy of unfolding (32,34,35) at the transition temperatures. The single electron reduction of $\underline{E}$. $\underline{coli}$ cytochrome b_{562} results in a 4.6 Kcal/mole change in free energy of unfolding at its Tm (7). In terms of Tm changes, single site mutations often have effects of several degrees to tens of degrees (9,35,36).

Computational Approaches

Computational approaches to studies of thermostability tend to be restricted by current technology and theory to Type I events. It is important to recognize the necessity for good experimental data and for removal of the effects of irreversible components of denaturation from such experiments before they can be used to test, verify or refute models developed computationally. This paper discusses computational approaches that may be useful in developing an understanding of extreme thermostability in proteins. These include analysis of seqeuence database information, the structures of mutant proteins, homology modeling, molecular mechanics and dynamics calculations, and free energies of mutational changes base on free energy perturbation calculations.

Analysis of Databases. The potential of analysis of database information for known amino acid compositions, sequences or X-ray structures to reveal the key to termostability has not yet been realized. Several well founded thermostability analyses (eg. 25-27) are not useful for this purpose because they consider biological stability of proteins (protein half-life in vivo, optimum growth temperature, normal body temperature, etc.). These kinds of comparisons do not separate Type I and Type II stabilities or distinguish between the contributions of denaturation, protein synthesis rates and enzyme catalyzed degradation to stability. Since all of these factors tend to vary from protein to protein and from organism to organism, the results of such studies do not make much contribution to our fundamental understanding of the role of composition, sequence or structure to the inherent stability of proteins.

Progress has been slow toward a rational understanding of protein folding and therefore protein stability based on compositions or sequences. For recent reviews see the books edited by   Nall and Dill (37) and Georgiou and de Bernardez-Clark (38). Some fairly consistent correlations have been observed between amino acid composition and thermostability of some proteins, and between amino acid sequence information, folding tendencies and stability.

Several attempts have been made to correlate thermostability with indices used to predict protein folding motifs and other structural properties (*22-24,39*).

Many experimental and theoretical studies have focused on the folding and stability of proteins and peptides that contain alpha helices.   The conventional wisdom is that folding of such proteins involves the formation of local helical segments fairly early in the folding process which promote the subsequent rearrangements that lead to the complete three dimensional structure (*2,3*).   Indeed several model peptides have been found to spontaneously form helices in aqueous solution.   These include the S-peptide from the N-terminal sequence of ribonuclease A (*40*), the Pα5 peptide from bovine pancreatic trypsin inhibitor (*41*) and some synthetic peptides based on these and other sequences (see *2, 39*).   One limitation of many of the shorter peptides for studies of stability is their relatively low helical content in solution and low thermal stability (*39*). These workers have recently shown that helicity in short peptides can be predicted with reasonable accuracy using a modified Zimm-Bragg model (*42*). This model uses position dependent, helix-coil stability parameters specific to each amino acid residue to calculate helical content of peptide sequences. Their study of ribonuclease A C- and S-peptide fragment analogs and alanine based peptides shows good agreement between calculated helicity and measurement. The model also makes resonable estimates of the effects of pH and temperature.

The data on composition and sequence differences between mesophilic and thermophilic proteins is not yet sufficient to draw any firm conclusions. However, there does not appear to be any obvious compositional or sequence characteristic   commonly   associated   with   temperature   stability (*15,19,21,26,36,43-46*).   Indeed, thermophilic and extremely thermophilic proteins are so much like their mesophilic counter parts that no generalizations have emerged about sequence differences between them.

Analysis of Mutant Structures. Analysis of the three dimensional structures from natural mutants and random mutagenesis (see *9,33*) tends to support two general conclusions:

1) More stable mutants tend to involve surface residues.

2) Less stable mutants tend to involve interior residues.

Neither of these results is surprising and both are suspect as guides for rational design of protein stability because of the nature of the samples studied and of several dramatic exceptions.   First, since most natural mutations involve surface residues of proteins, they provide little information as to the possible stabilizing influences within the interior.   Second, single site mutations are typically generated by random mutagenesis experiments that do not tend to favor stabilizing internal mutations. Stabilizing internal mutations are best accommodated with multiple compensating changes made to relieve adverse steric or electrostatic effects. Even so, there is at least one dramatic example

where a major change in thermostability was engineered by single site changes
in the interior of the protein. Zulli et al. (36), using lactate dehydrogenase,
changed interior threonine and serine residues to alanine with increases in the
the Tm of around 15°C.

Homology Modeling of Thermophilic Proteins. The purification and sequence
analyses of a variety of proteins from extreme thermophiles that have been
discussed in this book have generated considerable optimism for both
experimental and computational efforts toward understand thermostability.
The first step in a computational effort using molecular mechanics and
dynamics is to obtain an estimate of the 3-dimensional structure of such
proteins and of their less stable homologs. Obviously, the best approach would
be to have high-resolution X-ray or NMR structures for proteins of a
homologous series covering a range of thermostabilities. This information will,
no doubt, be available in time. Presently, however, the computational approach
must start with structures obtained by homology modeling of thermophilic
proteins using known structures of their less stable homologs. In several cases
discussed in these proceedings, extremely thermophilic proteins have been
described which show a high degree of sequence homology to less stable
proteins having known X-ray (or NMR) structures.

The X-ray structure database (47) contains several sets of structures of
closely related proteins. When such structures are examined in detail, it is
clear that there is considerable structural homology in them (48). Thus, a
straightforward way of modeling homologous proteins is suggested where known
structures are "mutated" computationally (see 48,49). Several specific
procedures have been described (50,51). The procedure used in our laboratory
involves two simple steps:

1)   The three dimensional structure of a homolog from the
     structure database is mutated into the unknown by editing
     its sequence substituting each differing amino-acid residue.

2)   The resulting "mutant" structure consisting primarily of
     atoms positioned as in the parent molecule is then
     subjected to energy minimization to relieve strain
     introduced by the editing process.

This method of homology modeling gives fairly accurate predictions of
structure when used for proteins of high amino-acid sequence homology (50,52).
The accuracy of the results depends to some extent on the method of
minimization used and on the details of the procedure. For example, in recent
studies modeling the known structures of rubredoxin, Stewart and collegues
(50,52) used the AMBER 2.0 and AMBER 3.0 force fields. The RMS deviation
in the positions of atoms for homology modeled structures were typically
within 1 Å of the corresponding minimized crystal structures. Indeed, when
the D. vulgaris structure was used to model the C. pasteurianum protein and

using an all-atom force field (see below), the modeled structures were both less than 0.6 Å RMS different from the minimized crystal structure.   This difference is actually less than the difference between the X-ray crystal structures as obtained from the Protein Data Bank before and after AMBER minimization.

The sequence of rubredoxin from the extreme thermophile P. furiosus exhibits strong sequence homology to all twelve of the mesophilic rubredoxins currently known (*21*).   Of the four rubredoxins of known three dimensional structure, those from D. gigas, D. vulgaris and C. pasteurianum show particularly high homology to that from P. furiosus (Wampler et al., manuscript in preparation).   This recent work has included homology modeling of the P. furiosus protein based on all three of these closely related homologs.   The results show that all three models of P. furiosus rubredoxin converge toward each other and away from the parent structures.   This can be shown quantitatively by some descriptive parameters from analysis of the minimized structures (Figure 2).   In all three cases the total AMBER energies and the

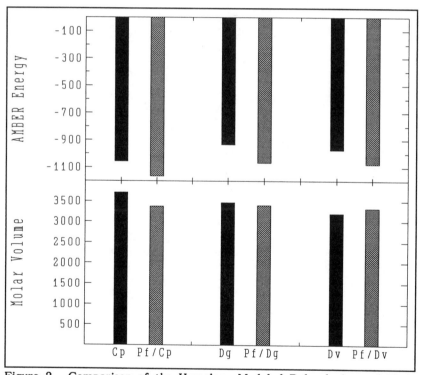

Figure 2.   Comparison of the Homology Modeled Rubredoxins and Their Parent Proteins.

molar volumes of the modeled P. furiosus rubredoxins (Pf/Cp, Pf/Dg and Pf/Dv in Figure 2) are nearly identical. On-the-other-hand, the values of the parent structures (Cp, Dg and Dv) differ from each other. It is easy to visualize this convergence when the structures are aligned in three dimensions (Figure 3).

The experience with known rubredoxin structures and the results from modeling the P. furiosus structure support the idea that homology modeling can be used to gain insight into the structures of thermostable homologs. As discussed below, these models can then be used in other computational efforts to help define the relationships between changes in amino acid sequence and thermostability.

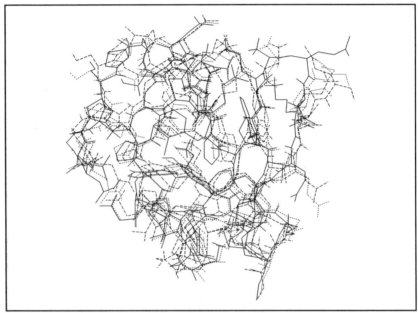

Figure 3. Aligned structures of Pyrococcus furiosus rubredoxin. Solid line = modeled from *C. pasteurianum*, Dotted line = from *D. gigas*; Dashed line = from *D. vulgaris*.

Molecular Mechanics and Molecular Dynamics Calculations. The basis of molecular mechanics and dynamics calculations (see 53, 54) that are used for modeling the interatomic interactions within and between proteins lies in the assumptions that nuclear and electronic motions can be treated separately (the Born-Oppenheimer approximation) and that, for practical purposes, average electron densities around each atom can be treated as point charges localized at the atomic centers. These assumptions make it possible to view the internal energy and the interatomic forces of a molecule or molecular system as a simple linear sum of interactions. Energy functions have then been derived (also called force fields) which present the problem as a simple linear sum of component energy contributions. In a typical situtation, the atomic parameters (bond distances, bond angles, point charges, etc.) needed to evaluate the energy are obtained from experimental values or quantum mechanical calculations. However, the parameters are also typically adjusted to give results consistent with structures in some test database.   Thus, the energies obtained are empirical.   With small molecules, the success of these approximations as implemented in the computer program MM-2 (*55*) is indicated by predictive accuracy which rivals experimental approaches (*56*).   For protein modeling, several programs have been described (*57,58,59*) which use slightly simpler equations for the interactions. Such simplifications are necessary in order to keep calculation times short enough for practical application of the procedure to problems with so many atoms (see *60*). The protein modeling program used in this laboratory is called AMBER (*61*). The AMBER force field consists of seven terms summed together as indicated in following equation.

Total Energy = Bond Energy + Bond-Angle Energy + Dihedral-Angle Energy +
Van-der-Waals Interactions + Electrostatic Interactions +
Hydrogen-Bonding Energy + Harmonic Constraints

The bond length and bond angle terms are represented in the calculation by simple harmonic potentials. The dihedral angle potential is cyclic with the number of minima per $360°$ rotation determined by the bond type. The nonbonding interactions within the protein are represented by a 6-12 van der Waals potential and by a simple coulombic electrostatic potential. Proper hydrogen bonding relationships are maintained with the aid of a separate 10-12 potential. Harmonic constraints can optionally be applied to restrict motion or changes in geometry. For the electrostatic interactions the dielectric constant used in the calculation can either be a constant or scaled by distance. Because the non-bonding calculations potentially involve all atom-by-atom pairings, this calculation is the most computationally intensive. Some reduction in time can be obtained by limiting the calculation of non-bonded interactions to only those atoms within some spherical radius about each atom (a cutoff). Other options allow further simplification.   For example, rather than representing the hydrogen atoms explicitly (an All Atom Force Field), the parameters for each heavy atom can be adjusted to simulate the effect of the attached hydrogens. This option is referred to as a United Atom force field. Even in this case, AMBER treats hydrogens on polar atoms explicitly. For speed, calculations are driven by lists of atoms involved for each term. For non-bonded interactions the list of interacting atoms changes as the distance between atoms changes (if a cutoff is involved).   This list can be updated with every complete calculation of the structure, but more typically it is only updated for every 10th or 100th calculation.   Solvation of the protein is simulated very approximately by the selection of the distance dependent dielectric constant, but it can be simulated more rigorously by inclusion of explicit solvating water where the molecular interactions of water are evaluated using a model function, TIP3P (62).

Molecular Mechanics Minimizations. For homology modeling and other applications where a low-energy structure is sought, the force field energy is minimized with respect to the x, y and z coordinates of the atoms. With so many variable parameters (the coordinates of all atoms), minimization can not be expected to find the global minimum. At best it will find a local minimum with a structure near to that of the initial coordinate set.

For molecular dynamics simulations, the empirical energy function is used to calculate forces on the atoms and Newton's equations of motion are solved for small time steps to calculate new atomic positions. A four step procedure is used (Figure 4): 1) the structure for dynamics simulation is minimized to relieve all unequal strain; 2) the force field is used to calculate accelerations to be applied to the atoms of the structure and it is "warmed" to the simulation temperature in steps; 3) the warmed structure is stabilized at temperature with removal of rotational and translational motion; 4) the simulation is continued and structures are periodically collected. The time line of Figure 4 illustrates the process. For warming, random velocities selected from an appropriate Boltzmann or Gaussian distribution of velocities for a given temperature are assigned to the atoms. Cycles of simulation are then applied to the structure where movements are calculated for short time intervals (typically 0.0005 ps) by application of these velocities and the accelerations due to the potential function. New positions are calculated for all of the atoms and the process is repeated for several cycles. Equilibration, like warming, involves periodic reassignment of the velocities with random velocities selected from the appropriate temperature distribution. Rotational and translational motion is removed. After equilibration, the simulation is continued and structures are collected. Temperature is maintained during the simulation by scaling velocities periodically.

Molecular mechanics alone does not offer much insight into the problems of protein stability as seen in the homology modeling results presented above (Figure 2). The scaled potential energy functions of molecular mechanics give resulting proteins close in conformation to the parent structures. While the three models of P. furiosus rubredoxin do seem to converge toward a common structure as mentioned above, there is no clear advantage in size or AMBER energy in the "thermostable" models.

Molecular Dynamics Simulations. Most theoretical work on protein stability using molecular dynamics has focused on the folding of small peptides. Levitt and coworkers have studied the folding of bovine pancreatic trypsin inhibitor (63,64) and carp myogen (65) using molecular mechanics with a considerably simplified representation of the rotational freedom of bonds in the residues and of the force field. Two recent studies have used molecular dynamics simulations to probe unfolding of small helical peptides. Brooks et

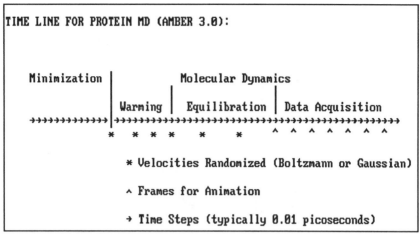

Figure 4. Molecular Dynamics Calculation Time Line. Periods are not to scale.

al. (*31*) used a fifteen residue polyvaline helix in their simulation studies using the simplified molecular mechanics approach described by Levitt (*63*) along with a Brownian dynamics component. With this simplified model they noted correlated, concerted transitions from helix to coil with unwinding rates at 298K limited to about 10^7/sec.

A more rigorous approach has been used with a short helical peptide like those from the experimental studies of Finkelstein et al. (*39*) (see above). Tirado-Rives and Jorgensen (*66*) homology modeled a 15 residue S-peptide fragment from the N-terminal of ribonuclease A. The sequence differs from the S-peptide in three residues (Lys1 -► Ala, Glu9 -► Leu and Gln11 -► Glu). These changes give a peptide with a relative high α-helical content (~45% at 276 K) and a melting temperature below 60°C (333 K). AMBER 3.0 molecular dynamics simulations of this peptide starting in the helical configuration were carried out in TIP3P model water (*62*). The results of three simulations at two different temperatures showed that much larger changes in the structure occur at high temperature (358 K) than during the simulation at low temperature (278 K). The RMS change in atomic positions increased to around 2 Å at 358 K compared to 1 Å RMS change at 278 K.   Detailed analysis of the trajectories shows that the helical content during two high temperature (358 K) simulations was consistently lower and the structure was more extended. It is important to note that this difference between high and low temperature simulations was observed after a much shorter time and with a much more complete force field than was used by Brooks et al. (*31*). While the transition observed is not completely helix to coil, helical content does decrease at the higher temperature as measured by hydrogen bonding patterns and by a tendency toward a more extended, 3,10 helical conformation.

As with the Tirado-Rives and Jorgensen (*66*) study, our studies of P. furiosus rubredoxin also indicate that changes in amino acid composition change the molecular dynamic behavior of homology models in a way that is consistent with increased thermostability (Wampler et al., manuscript in preparation). The models described above (Figure 3) and their parent minimized X-ray crystal structures were subjected to molecular dynamics simulations at relatively high temperatures. At 70°C in all cases the total energies of the "thermostable" models tended to be lower than those of the parent proteins and the molar volumes revealed less expansion (Figure 5). At a higher temperature (140°C)

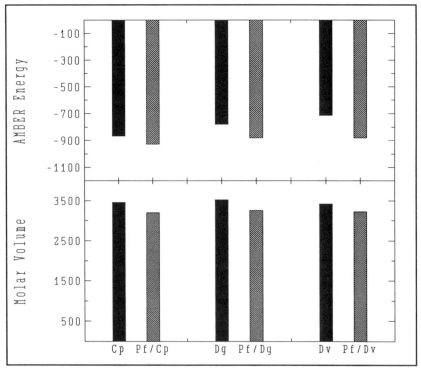

Figure 5.   Comparison of the 70°C Simulation Results for Modeled Rubredoxins and Their Parent Proteins.

that is presumably above the Tm of all four proteins, the AMBER energies and volumes of the P. furiosus rubredoxin models rise (Figure 6). In two cases (Cp and Dg) the parent structures do not change as much (relative to 70°C) and have average parameters more nearly like those of their corresponding homology models. The AMBER energy of the D. vulgaris structure, on-the-other-hand, continues to rise (less stable). Together these observations suggest

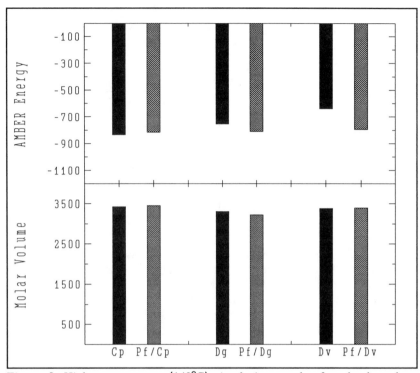

Figure 6. High temperature (140°C) simulation results for the homology modeled rubredoxins and their parent structures.

that the changes in amino acid sequence used to generate the homology models of the P̱. furiosus protein do stabilize its structure at 70°C relative to the less-thermophilic structures. However, at 140°C all of the structures are less stable and more expanded. These results seem to support the use of homology modeling and molecular dynamics simulations for analyzing stabilizing changes in amino acid sequence.

Free Energy Perturbation Calculations and Protein Stability.   Another indication that molecular mechanics force fields hold promise for understanding thermostability is from free energy difference calculations based on the thermodynamics cycle represented by Figure 7. The effect of a mutation on the relative stability of a structure is evaluated using molecular dynamics simulations to obtain an ensemble average of the energy change effected by perturbation of the structure from one containing the wild-type residue to one containing the mutant residue. For a detailed discussion of the procedures and successes of approaches to calculate free energies using simulations see the book edited by van Gunsteren and Weiner (*67*). The typical calculation is summarized by Figure 8. The parameter λ controls the "mutational" event.

When λ is zero, the structure is the wild-type. When $\lambda=1$, the structure has the mutant residue. At intermediate values, each structure makes a fractional contribution to the calculation.

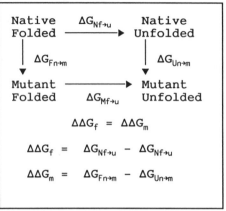

Two such calculations are involved in defining the free energy difference in stability between the native and mutant forms of a protein. $\Delta G_{Fn \to m}$ is calculated by the procedure of Figure 8 using the native structure. The second calculation, $\Delta G_{Un \to m}$, requires a model for the unfolded structure. With a complete sequence of a large protein or peptide, there are too many possible structures for the random coil of the unfolded state to be able to obtain a complete ensemble from a single simulation. The typical solutions used in the examples discussed below are for the unfolded state to be modeled by an extended conformation of a short peptide homologous

Figure 7.   Free Energy Cycle for Unfolding a Native and Mutant Protein. Symbols: n=Native, m=Mutant, f=Folded, u=Unfolded & G=Free Energy.

Relative Free Energy From MD:

For the change A→B

$$\Delta G_{A \to B} = \int_{\lambda = 0}^{1} \left\langle \delta E_{(I,X,Y,Z)} \middle/ \delta \lambda \right\rangle_{\lambda} d\lambda$$

where ⟨...⟩ is the ensemble avg.

and λ controls the transformation.

Figure 8. Equation Used for Free Energy Calculation using the Perturbation Method.

to the region of the mutation. Given both values for free energies of the "mutational" changes in the folded and unfolded state, the free energy difference between them ($\Delta\Delta G_m = \Delta G_{Fn \to m} - \Delta G_{Un \to m}$) is equal to the free energy difference for stability due to the mutation ($\Delta\Delta G_f = \Delta G_{Nf \to u} - \Delta G_{Nf \to u}$). Thus, the first law of thermodynamics (energy conservation), allows us to calculate the difference that can not be simulated using differences that can be simulated. Obviously, simulation of the full unfolding process which experimentally takes periods of time up to milliseconds or even seconds is not accessible to our current computer technology. For example, a recent simulation of protein dynamics of a small pair of proteins required 86 hours of supercomputer time to simulate 65 ps of motion (*60*).

Free energy perturbation calculations have been applied to two different protein structures to evaluate the effect of mutations. With calbindin D_{9k} (*68*) a simple force field of coulombic energies was used to evaluate three different changes (Table II). While the accuracy of the results is not impressive, the order of the predicted effect (and its sign) correlated well with experimental values. Similarly, two changes in T4 Lysozyme were simulated (*69,70*) using more complete force fields. The results in both cases are of the same magnitude and sign as the experimental measurements (Table II). As with the simpler dynamics results discussed above, these result are encouraging and suggest that computational procedures based on molecular mechanics force fields for proteins may be an aid in understanding thermostability.

Table II. Free Energy Results for Single Site Mutations in Proteins*

Protein	Potential Function	Method	Unfolded Model	pH	Mutation	$\Delta\Delta G$ (Kcal) calc.	expt.
Calbindin D9k	Coulombic[1]	MUMOD[2]	freely rotating chain	7.0	E17→Q D19→N E26→Q	4.7 7.6 1.4	1.6 3.0 0.4
Lysozyme (T4)	CHARMM[3]	CHARMM[3]	7 residue fragment	5-6	R96→H	-1.6	-2.8
Lysozyme (T4)	AMBER[4]	AMBER[4]	3 residue fragment	?	T157→V	1.9	1.6

* References for calculations: Calbindin (*68*), Lysozyme using CHARMM (*69*), Lysozyme using AMBER (*70*)
[1] see reference *71*
[2] see reference *57*
[3] see reference *58*
[4] see reference *59*

Conclusions

Computational chemistry and computer-based analyses hold considerable promise to help define and dissect the problem of extreme thermostability of proteins from the extraordinary organisms which have been the focus of this book. As more detailed sequence and structure information becomes available for this class of proteins, much more definitive statistical analyses of compositions or sequences are likely to be developed. Molecular modeling and computational chemistry approaches may indeed give insight into the detailed interatomic interactions that furnish the driving forces that make thermostable proteins more rigid and resistent to unfolding. At minimum these approaches can be used to guide experimental work by investigating potentially important interactions that can then be analyzed using molecular biology techniques. Molecular dynamics and free energy perturbation calculations can be performed at unrealistic temperatures experimentally, thus probing and compressing the time requirements for the calculation and cleanly separating the physical and chemical events of the complicated processes of thermal denaturation. The computational chemistry results to date seem to give a consistent picture and one that correlates fairly well with the known experimental data. However, they also challenge the experimentalist to obtain more rigorous measurements of stability, melting temperatures, calorimetric data and biophysical measurements that cleanly separate the reversible and irreversible phenomena.

Acknowledgments

The work from this laboratory described in this paper was supported by the National Institutes of Health (Grant No. GM41482), the North Atlantic Treaty Organization (Grant No. 0404/88), the National Science Foundation (Training Grant No. DIR-9014281) and the University of Georgia Computing and Network Services. Special thanks are due to Dr. Walter McCrae, Assistant Vice President for Computing, for support funds and computer time on the high performance machines.

Literature Cited

1. Privalov, P. L., and Gill, *Advances in Protein Chemistry* **1988** 39, 191-234.
2. Baldwin, R. L. *Trends Biochem. Sci.* **1989** 14, 291-294.
3. Montelione, G. T., and Scheraga, H. A. *Acc. Chem. Res* **1989** 22, 70-76.
4. Ptitsyn, O. B., and Semisotnov, G. V. in *Conformatons and Forces in Protein Folding*; Nall, B. T., and Dill, K. A., Eds; American Association for the Advancement of Science, Washington, D. C., **1991**; pp. 155-168.
5. Wetzel, R., Perry, L.J., Baase, W. A., and Becktel, W. J. *Proc. Natl. Acad. Sci., USA* **1988** 85, 401-405
6. Sagar, S. L., and Domach, M. M. in *Protein Folding*; Georgiou, G., and de Bernardez-Clark, E., Eds; American Chemical Society, Washington, D. C., **1991**; pp. 64-78.
7. Fisher, Mark T. *Biochemistry* **1991** 30, 10012-10018.
8. Ahern, T. J., and Klibanov, A. M. in *Protein Structure, Folding and Design*; D. L. Oxender, Ed; Alan R. Liss, Inc., New York, **1986**; pp. 283-289.
9. Manning, M. C., Patel, K., Borchardt, R. T. *Pharmaceutical Research* **1989** 6, 903-918.
10. Ahern, T. J., and Klibanov, A. M. *Science* **1985** 228, 1280-1284.

11. Ahern, T. J., Casal, J. I., Petsko, G. A., and Klibanov, A. M. *Proc. Natl. Acad. Sci. USA* **1987** 84, 675-679.
12. Pace, C. N., Shirley, B. A., and Thomson, J. A. in *Protein Structure, A Practical Approach*; T. E. Creighton, Ed.; IRL Press, Oxford University Press, Oxford, **1989** pp. 311-330.
13. Daniel, R. M. in *Protein Structure, Folding and Design*; Oxender, D. L., Ed.; Alan R. Liss, Inc., New York, **1986** pp. 291-296.
14. Costantino, H. R., Brown, S. H., and Kelly, R. M. *J. Bateriol.* **1990** 172, 3654-3660.
15. Zwickl, P., Fabry, S., Bogedain, C., Haas, A., and Hensel, R. *J. Bacteriol.* **1990** 172, 4329-4338.
16. Aono, S., Bryant, F. O., and Adams, M. W. W. *J. Bacteriol.* **1989** 171, 3433-3439.
17. Bryant, F. O., and Adams, M. W. W. *J. Biol. Chem.* **1989** 264, 5070-5079.
18. Saiki, T., Iijima, S., Tohda, K., Beppu, T., and Arima, K. in *Biochemistry of Thermophily*; Friedman, S. M., Ed.; Academic Press, New York, **1978** pp. 287-303.
19. Nojima, H., Ikai, A., Nota, H., Hon-nami, K., and Oshima, T. in *Biochemistry of Thermophily*; Friedman, S. M., Ed.; Academic Press, New York, **1978** pp. 305-323.
20. Blumentals, I. I., Bobinson, A. S., and Kelly, R. M. *Appl. Environ. Microbiol.* **1990** 56, 1992-1998.
21. Blake, P. R., Park, J-B., Bryant, F. O., Shigetoshi, A., Magnuson, J. K., Eculston, E., Howard, J. B., Summers, M. F., and Adams, W. W. *Biochemistry* **1991** 30, 10885-10895.
22. Bull, H.B. and Breese, K. *Archives of Biochemistry and Biophysics* **1973** 158, 681-686.
23. Stellwagen, E. and Wilgus, H. *Nature* **1978** 275, 342-343.
24. Stellwagen, E. and Wilgus, H. in *Biochemistry of Thermophily*; Friedman, S. M., Ed.; Academic Press, New York, **1978** pp.223-232.
25. Guruprasad, K., Reddy, B.V.B., and Pandit, M.W. *Protein Engineering* **1990** 4, 155-161.
26. Argos, P., Rossmann, M.G., Grau, U.M., Zuber, H., Frank, G., and Tratschin, J.D. *Biochemistry* **1979** 18, 5698-5703.
27. Ljungdahl, L. G., and Sherod, D. in *Extreme Environments: Mechanisms of Microbial Adaptation*; M. R. Heinrich, Ed.; Academic Press, N. Y., **1976** pp. 147-187.
28. Chen, B.-L., and King, J. in *Protein Folding*; Georgiou, G., and de Bernardez-Clark, E., Eds; American Chemical Society, Washington, D. C., **1991** pp. 119-152.
29. Papavassiliou, P., and Hatchikian, E. C. *Biochim. Biophys. Acta* **1985** 810, 1-11.
30. Dill, K. A., and Alonso, D. O. V. in *Protein Structure and Protein Engineering*; Winnacker, E.-L., and Huber, R., Eds; Springer-Verlag, Berlin, **1988** pp. 51-58.

31. Brooks, C. L., III, Karplus, M., and Pettitt, B. M. *Adv. Chem. Phys.* **1988** 71, p. 128-129.
32. Pace, C.N. *Tibs* **1990** 15, 14-17.
33. Abler, T. *Annu. Rev. Biochem* **1989** 58, 765-798.
34. Connelly, P., Ghodaini, L., Hu, C-Q., Kitamura, S., Tanaka, A., and Sturtevant, J.M. *Biochemistry* **1991** 30, 1887-1891.
35. Nicholson, H., Anderson, D.E., Dao-pin, S., and Mathews, B.W. *Biochemistry* **1991** 30, 9816-9828.
36. Zulli, F., Schneiter, R., Urfer, R., and Zuber, H. *Bilo. Chem. Hoppe-Seyler* **1991** 372, 363-372.
37. *Conformations and Forces in Protein Folding*; Nall, B. T., and Dill, K. A., Eds; American Association for the Advancement of Science, Washington, D. C., **1991**
38. Georgiou, G., and de Bernardez-Clark, E. *Protein Folding*, American Chemical Society, Washington, D. C., **1991**
39. Finkelstein, A.V., Badretdinov, A.Y., and Ptitsyn, O.B. *Proteins* **1991** 10, 287-299.
40. Klee, W. A. *Biochemistry* **1968** 7, 2731-2736
41. Goodman, E. M., and Kim, P. S. *Biochemistry* **1989** 28, 4343-4347
42. Zimm, B. H., and Bragg, J. R. *J. Chem. Phys.* **1959** 32, 526-535.
43. Zuber, H. in *Biochemistry of Thermophily*; Friedman, S. M., Ed.; Academic Press, Newy York, **1978** pp. 267-285.
44. Kristjansson, M.M., and Kinsella, J.E. *Int. J. Peptide Protein Res.* **1990** 36:2, 201-207.
45. Wilson, K. S., Vorgias, C. E., Tanaka, I., White, S. W., and Kimura, M. *Protein Engineering* **1990** 4, 11-22.
46. Fontana, A. in *Life Under Extreme Conditions*; G. di Prisco, Ed.; Springer-Verlag, Berlin, **1991** pp. 89-113.
47. Bernstein, F. C., Koetzle, T. F., Williams, G. J. B., Meyer, E. F., Brice, M. D., Rogers, J. B., Kennard, D., Shimanouchi, T., and Tasumi, M. *J. Mol. Biol.* **1977** 112, 535-542.
48. Chothia, C., and Lesk, A. M. in *Computer Graphics and Molecular Modeling*; R. Fletterick, Ed.; Cold Spring Harbor Symposium, Cold Spring Harbor, N. Y., **1986** pp. 33-37.
49. Jones, T. A., and Thirup, S. *EMBO J.* **1986** 5, 819-822.
50. Stewart, D. E., Weiner, P. K., and Wampler, J.E. *J. Mol. Graph.* **1987** 5, 137-144.
51. Sali, A., Overington, J.P., Johnson, M.S., and Blundell, T.L. *Tibs* **1990** 15, 235-240.
52. Stewart, D. E., *The Structure, Interactions, and Dynamics of Electron Transport Proteins from Desulfobivrio: A Molecular Modeling and Computational Study*; Ph. D. Dissertation, University of Georgia, Athens, GA., **1989**
53. Burkert, U., and Allinger, N. L. *Molecular Mechanics*; American Chemical Society, Washington, D. C., **1982**
54. van Gunsteren, W.F. *Protein Engineering* **1988** 2, 5-13.

55. Allinger, N. L. *J. Am. Chem. Soc.* 1977 <u>99</u>, 8127-8134.
56. Engler, E. M., Andose, J. E., and Schleyer, P. von R. *J. Am. Chem. Soc.* 1973 <u>95</u>, 805-8025.
57. Teleman, O., and Jonsson, B. *J. Comput. Chem.* 1986 <u>7</u>, 58-66.
58. Brooks, B. R., Bruccoleri, R. E., Olafson, B. D., States, D. J., Swaminathan, S., and Karplus, M. *J. Comput. Chem.* 1983 <u>4</u>, 187-217.
59. Singh, U. C., Wiener, P. K., Caldwell, J. W., Kollman, P. A. *AMBER (USCF), Version 3.0*; Department of Pharmaceutical Chemistry, University of California, CA., **1986**
60. Wampler, J. E., Stewart, D. E., and Gallion, S. L. in *Recent Developments in Computer Simulation Studies in Condensed Matter Physics*; D. Landau, Ed.; Springer-Verlag, Heidelberg, **1990** pp. 68-84.
61. Weiner, S. J., Kollman, P. A., Chase, D. A., Singh, U. C., Ghio, C., Alagona, G., Profeta, S., and Weiner, P. *J. Am. Chem. Soc.* 1984 <u>106</u>, 765-784.
62. Jorgensen, W. L., Chandrasekhar, J., Madura, J. D., Impey, R. W., and Klein, M. L. *J. Chem. Phys* 1983 <u>79</u>, 926.
63. Levitt, M. *J. Mol. Biol.* 1976 <u>104</u>, 59-107
64. Warshel, A., and Levitt, M. *Nature* 1975 <u>253</u>, 694-698
65. Warshel, A., and Levitt, M. *J. Mol. Biol.* 1976 <u>106</u>, 421-437
66. Tirado-Rives, J. and Jorgenson, W.L. *Biochemistry* 1991 <u>30</u>, 3864-3871.
67. Van Gunsteren, W. F., and Weiner, P. K., Eds. *Computer Simulation of Biomolecular Systems*; ESCOM Science Publishers B. V., Leiden, The Netherlands, **1989**
68. Akke, M., and Forsen, S. *Proteins* **1990** <u>8</u>, 23-29.
69. Tidor, B. and Karplus, M. *Biochemistry* **1991** <u>30</u>, 3217-3228.
70. Dang, L. X., Merz, K. M., and Kollman, P.A. *J. Am. Chem. Soc.* 1989 <u>111</u>, 8505-8508.
71. Ahlstrom, P., Teleman, O., Jonsson, B., and Forsen, S. *J. Am. Chem. Soc.* 1987 <u>109</u>, 1541-1551.

RECEIVED January 15, 1992

Chapter 12

DNA-Binding Proteins and Genome Topology in Thermophilic Prokaryotes

D. R. Musgrave[1], K. M. Sandman[2], D. Stroup[2], and J. N. Reeve[2]

[1]Department of Biological Sciences, The University of Waikato, Private Bag 3105, Hamilton, New Zealand
[2]Department of Microbiology, The Ohio State University, Columbus, OH 43210

Thermophilic microorganisms must protect their genomes from heat denaturation while allowing structural changes to occur, including DNA strand separation, during gene expression, DNA replication, repair and recombination. In this chapter we review genome topology and describe the formation of novel nucleosome-like structures containing DNA molecules constrained in positive toroidal supercoils by histone-related proteins isolated from thermophilic methanogens. The DNA binding properties of these proteins *in vitro* and their abundance *in vivo* suggest that they play roles in the architecture and protection of the genomes of these thermophilic *Archaea* from heat denaturation. Hyperthermophiles, including the methanogen *Methanothermus fervidus*, also contain reverse gyrase. The participation of this positive supercoiling topoisomerase and the role of a high intracellular salt concentration in providing heat resistance are also discussed.

Bacterial Genome Structure and DNA Binding Proteins

Both genetic and physical methods have demonstrated that the *E. coli* chromosome is a single, circular, double-stranded DNA (dsDNA) molecule, containing ~4.7 Mbp, approximately 1 mm in length (*1-4*). This chromosome is therefore ~1000X longer than the cell into which it is packaged, and allowing for the fact that the chromosomal nucleoid does not totally fill the cell, an intracellular packing density of between 14 and 34 mg DNA/ml has been calculated (*5,6*). Analyses of *E. coli* nucleoids produced using high counterion concentrations have revealed that the chromosome is organized into topologically-independent, looped domains, each under negative torsional stress (*7-10*). Although the details of chromosomal packaging in *E. coli* have yet to be determined, several small basic histone-like proteins, such as HU, which can constrain DNA in negative toroidal supercoils *in vitro*, are known to be present in relatively large amounts and bead-like structures resembling nucleosomes have been visualized by electron microscopy (*10-12*). HU also plays a role in facilitating the interaction of other proteins with DNA, apparently by altering the topology or flexibility of the DNA (*13-16*). In agreement with a structural role for HU, its DNA binding is not sequence specific, however its concentration at ~60,000 monomers per exponentially growing *E. coli* cell appears insufficient to contribute maximally either to toroidal supercoiling *in vivo* or to its accessory roles (*10,12,17,18*). This and the rapid dissociation of HU/DNA complexes *in vitro* (*11*) has led to the term "transitional DNA coiling" (*18*) to indicate that HU/DNA interactions must be dynamic and balanced at an equilibrium *in vivo* determined by interactions with other DNA binding proteins and by the DNA structures which form due to these interactions. Transitional DNA coiling may thus contribute to gene regulation in addition to providing the DNA condensation needed to account for the observation that only ~50% of the linking deficit in isolated *E. coli* nucleoids is found as plectonemic supercoils (*10,12,18-20*). HU binds to repetitive, extragenic palindromic (REP) sequences found in the *E. coli* chromosome, and in doing so stimulates DNA gyrase binding to these sequences. Yang and Ames have proposed therefore (*21*) that the individual supercoiled domains in the *E. coli* genome are anchored to each other or to the membrane by a REP/HU/DNA gyrase complex which functions in a manner analogous to the role proposed for topoisomerase II in eucaryal chromatin (*22*). The viability of HU deficient *hupA-hupB* double mutants clearly demonstrates that HU is not essential for growth in the laboratory at 37°C (*23*). Under these conditions HU functions are presumably undertaken by one or more of the other histone-like proteins known to be present in *E. coli* (*10,12,24*). HU deficiency does cause temperature sensitivity and HU protein is essential for Mu phage development and the replication of some plasmids (*25,26*). The location of HU *in vivo* is still a matter of conjecture. *In situ* immuno-gold labeling experiments, using anti-HU antibodies, have indicated that HU is not associated with the majority of the genomic DNA but rather is located at the periphery of the nucleoid, possibly associated with RNA polymerase and topoisomerase I (*5,27*). In contrast, fluorescent HU proteins, introduced into permeabilized but still viable *E. coli* cells, bind *in vivo* throughout the *E. coli* nucleoid (*28*).

Eucaryal Nucleosomes

In *Eucarya* the basic packaging unit of chromatin is the nucleosome in which 146 bp of DNA are wrapped 1.75 times around a protein core formed from an octamer of histone proteins (*29,30*). The shallow path taken by the DNA molecule as it wraps around the core of histones requires significant DNA bending and DNA sequences which have the capacity to bend have been proposed as major determinants of nucleosome positioning (*31,32*). Overall the helical periodicity of DNA molecules constrained in nucleosomes is decreased from the ~10.6 bp normally found for DNA molecules in solution (*33,34*) to 10.2 bp (*35*), although this is an average value as within a nucleosome there are DNA regions with 10.7

bp/turn and regions with 10 bp/turn (*36*). This net overwinding of the DNA partly resolves the paradox that the assembly of a nucleosome results in a linking number change of only one, significantly less than the 1.75 predicted from the number of superhelical turns of DNA in a nucleosome (*37*). In nucleosomes the DNA is wrapped in a left-handed, negative toroidal supercoil but is not otherwise torsionally stressed.   As negative toroidal wrapping is favored thermodynamically on negatively supercoiled DNA, nucleosomes transfer preferentially from positively supercoiled to negatively supercoiled DNAs (*39*). The positive supercoiling which occurs in front of a transcription complex has been proposed to facilitate the removal of nucleosomes from this location.   The negative supercoiling which accumulates behind the transcription complex is presumed then to direct their reassembly (*38,39*).

Sequence-independent eucaryal DNA binding proteins contain amino acid sequences which form positively-charged DNA binding motifs that recognize DNA conformation rather than DNA sequences (40).   These motifs appear to interact with the minor groove in dsDNA regions containing five or more A:T bp, presumably facilitated by the absence of the 2 amino-group which protrudes into the minor groove of G:C bp and by the size, shape, charge distribution and hydrogen bonding potential of A:T regions.   DNA binding by synthetic oligopeptides which contain these motifs indicates that conformational rigidity of the DNA binding domain *per se* is not required and that it is apparently the flexible nature of these proteins that allows them to bind to slightly different DNA sequences.   Proteins with greater structural rigidity and increased sequence specificity are needed to fulfil other cellular functions (*32,40*).

Archaeal Genome Structure and DNA Binding Proteins

DNA-DNA hybridization and pulse field gel electrophoresis have revealed a size distribution for archaeal genomes similar to that of *Bacteria*, ranging from 8.4 x 10^8 Da (1.27 Mbp) for *Thermoplasma acidophilum* to 2.5 x 10^9 Da (3.48 Mbp) for *Halococcus morrhuae* (*41*). Because the cellular dimensions are approximately the same, the extent of DNA condensation required for nucleoid packaging in the *Archaea* must be similar to that needed in the *Bacteria*.   DNA binding proteins have been isolated from species in all phylogenetic branches of the archaeal domain (*41*) and, in some cases, macromolecular structures reminiscent of eucaryal nucleosomes have been observed (*42-44*).   The DNA binding proteins HTa from *Thermoplasma acidophilum* (*42,45*), HMf from *Methanothermus fervidus* (*44*) and MC1 from *Methanosarcina* species (*46*) are well characterized and provide significantly different examples.   The amino acid sequence of HTa has regions similar to both HU and eucaryal histones (*47*) which led to the suggestion that it may be a link between the *Bacteria* and *Eucarya* (*42,43,47*). Functionally, HTa behaves like a eucaryal histone, remaining bound to DNA during isolation, resisting dissociation at relatively high ionic strengths and condensing DNA *in vitro* into small uniform globular particles.   HTa and HMf are both very effective in protecting dsDNA molecules from heat denaturation (*42,48*). HMf preparations contain two very similar, small polypeptides designated HMf-1 and HMf-2.

Determination of the DNA sequence of *hmfB*, the gene encoding HMf-2 has revealed that ~30% of the amino acid residues in HMf-2 are conserved in the consensus sequence derived for eucaryal histones H2A, H2B, H3 and H4. HMf appears, therefore, to be a histone *(44)*. HMf binding to dsDNA *in vitro* also results in compact structures which visibly resemble eucaryal nucleosomes. The sequence of MC1 is not related to HTa, HMf, HU or to histones and this protein does not form globular structures with DNA *in vitro* although it can provide heat resistance *(43,46)*. DNA molecules are constrained in negative toroidal supercoils in the eucaryal nucleosome, the bacterial (HU) DNA complex and when bound by MC1. However, as described in detail below, DNA molecules in the nucleosome-like structures formed by HMf are constrained in positive toroidal supercoils with an increased helical periodicity of 11.0bp (88).

Topoisomerases

Since the first description of a topoisomerase in 1971 *(49)*, the number of functionally distinct topoisomerases has continued to grow *(50)*. The number of cellular events thought to be regulated by changes in DNA supercoiling or by DNA structural motifs such as cruciforms, H-form triplexes or Z-DNA resulting from the torsional stresses introduced by topoisomerases, has increased in parallel *(12,24,50-52)*. In *E. coli* it has been argued that the opposing activities of DNA gyrase (DNA topoisomerase II), which can introduce negative supercoils at the expense of ATP, and topoisomerase I, which relaxes negatively supercoiled DNA, might determine the overall intracellular level of chromosomal supercoiling *(51-54)*. As an alternative, Liu and Wang *(38)* have proposed a model in which positively and negatively supercoiled domains are generated predominantly by transcription and replication complexes tracking along the DNA molecule, positive supercoils being produced ahead of these complexes and negative supercoils behind. These two proposals for establishing the overall genomic superhelicity *in vivo* differ primarily in the role ascribed to DNA gyrase. The Liu and Wang proposal *(38)* employs DNA gyrase to remove positive supercoils produced ahead of the transcription complex rather than to actively introduce negative supercoils. Since a positive toroidal supercoil is produced in the substrate DNA molecule as it wraps around DNA gyrase before catalysis *(55,56)*, this enzyme should interact preferentially with positively supercoiled DNA.   Such an interaction would be favored thermodynamically. The *E. coli* topoisomerase I, on the other hand, should interact preferentially with negatively supercoiled DNA because it binds more readily to single stranded DNA (ssDNA) which is more likely to be present in negatively supercoiled regions of duplex DNA *(49,57,58)*.

The overall superhelicity of DNA molecules *in vivo* will also be affected by temperature, intracellular ionic strength and pH. For example, the helix winding angle per base changes *in vitro* by between -0.012° and -0.014° ±0.001° for each °C change; the negative sign indicating that this angle decreases (unwinding) with increasing temperature *(59,60)*. In *E. coli* the overall superhelicity of the genome has been reported to be regulated by the ATP/ADP ratio and exogenously added salt may cause a change in genomic superhelicity indirectly by changing the intracellular ATP/ADP ratio (61). Changes in the intracellular ionic strength may however play a more direct role in genome structure in some thermophiles as described below.

Thermophily and Reverse Gyrase

The discovery of a novel topoisomerase, named reverse gyrase, in *Sulfolobus acidocaldarius* that can introduce positive supercoils into closed circular DNA molecules at the expense of ATP suggested that positive supercoiling might protect the genome of this thermophile from heat denaturation (*62*). Positively supercoiled DNA has a helical periodicity of greater than 10.6 bp and should be intrinsically more resistant to heat denaturation (*63*). Surveys of *Bacteria* and *Archaea* for reverse gyrase have now documented this activity in the hyperthermophilic archaeal species *Pyrodictium occultum, Pyrobacculum islandicum, Pyrococcus furiosus, Archeaoglobus fulgidus, Methanopyrus kandleri, Acidianus infernus* and *Methanothermus fervidus* (*64*), in the hyperthermophilic bacterial species *Thermatoga* (*65*), and in several novel sulfur metabolizing *Archaea*, all with optimal growth temperatures above 75°C (*66*). The presence of reverse gyrase appears to correlate with growth above 75°C, as related species growing below this temperature do not contain this activity. Reverse gyrases have been purified to homogeneity and characterized from *Sulfolobus acidocaldarius* (*67*), *Sulfolobus solfataricus* (*68*), *Desulfurococcus amylolyticus* (*69*) and strain AN1, a *Thermococcus*-like isolate (*68*). These enzymes are all single, large polypeptides with molecular masses ranging from 120 to 135 KDa. They have pH optima of approximately 6.5 at 75°C, ionic requirements of 150 to 250 mM NaCl or KCl and 10mM $MgCl_2$ and energy requirements of $> 10\mu M$ ATP. Mechanistically, they are all similar to bacterial type I topoisomerases in preferring single stranded regions of duplex DNA molecules and having a similar substrate and cleavage site specificity of 5'--CNNN* ---3' (*63,67-70*). Binding of reverse gyrase to negatively supercoiled templates should also be favored by the substrate DNA molecules being wrapped around this enzyme in a negative toroidal supercoil before catalysis. However, this property of reverse gyrase has been detected so far only with inactive molecules of the enzyme (*71,72*). Reverse gyrases, like eucaryal topoisomerase I, may also interact preferentially with nucleotide sequences which are intrinsically bent (*73*). Although all reverse gyrases appear structurally and catalytically similar, common antigens were not detected in other hyperthermophilic *Archaea* by antibodies raised against the reverse gyrase from *S. acidocaldarius* (*68*). The only firm evidence for positively supercoiled DNA *in vivo* is that the encapsidated genomes of the virus-like particle SSVI released by *Sulfolobus acidocaldarius* (*74*) are positively supercoiled. The lack of plasmids in the organisms studied to date that contain reverse gyrase has precluded further analysis of extrachromosomal supercoiling and technical difficulties arising from the cellular architecture of *Sulfolobus* species have so far prevented the use of nucleoids from this species for analysis of genomic supercoiling (*75*).

Thermophily and a High Internal Salt Concentration

Hyperthermophiles must protect their genomes from heat denaturation. This is a particularly acute problem for microorganisms such as *Methanothermus fervidus* (33%mol G+C; Tm 82.8°C; optimum growth at 83°C) and *Pyrococcus woesei* (37.5%mol G+C; Tm 84.8°C; optimal growth at 97°C) (76,77) which have genomic G+C contents so low that their genomes would denature *in vitro*, in low ionic strength solutions, at temperatures below their optimal growth temperature (78). High salt concentrations prevent heat denaturation of dsDNA *in vitro* and may also be used *in vivo* by *M. fervidus*, *Methanothermus sociabilis* and *Methanobacterium thermoautotrophicum* to help stabilize their dsDNA genomes. These cells contain high levels of potassium 2',3'-cyclic diphosphoglycerate (K_3cDPG), reportedly as high as 1060mM K^+ and 320mM $cDPG^{3-}$ in *M. fervidus* (78). The concentration of these ions increases with increasing growth temperature, suggesting a temperature regulated synthesis of this compound and that it functions as a 'thermostabilizing' agent. As not all hyperthermophilic *Archaea* contain high intracellular salt concentrations this is clearly not a universal heat resistance mechanism and its use by thermophilic methanogens may reflect their relatively close phylogenetic relationship to archaeal halophiles (79). These methanogens may, in fact, not have retained their high internal salt concentrations primarily for genome stability but to provide increased heat resistance to their proteins. Enzymes purified from *M. fervidus* do have much longer functional stabilities *in vitro* in the presence of cDPG (78). DNA binding proteins, such as HMf, may then not be needed to hold the DNA strands of the genome together but rather to facilitate their separation in the presence of high internal salt concentrations (see below).

DNA Binding Proteins from Thermophilic *Methanobacteriales* that Cause Positive Toroidal Supercoiling

M. fervidus and *M. thermoautotrophicum* are related methanogenic *Archaea* on separate phylogenetic branches within the *Methanobacteriales* (79). *M. fervidus* grows optimally at 83°C and *M. thermoautotrophicum* at 65°C; they are designated as hyperthermophilic and thermophilic, respectively. *M. fervidus* contains the positive supercoiling topoisomerase reverse gyrase (64) but this activity has not been detected in *M. thermoautotrophicum* (64,68). The abundant DNA binding protein HMf, from *M. fervidus*, binds to plasmid molecules *in vitro* to produce positively supercoiled topoisomers at protein to DNA mass ratios above 0.3:1 and negatively supercoiled topoisomers at lower ratios [Figure 1 and (88)]. This suggests that either the DNA molecule wraps around cores of HMf in different directions at different protein to DNA ratios or that the positive and negative topoisomers observed result from different molecular interactions. Hydroxyl radical footprinting has revealed that DNA molecules, bound by HMf, at all protein to DNA mass ratios studied, have ~11bp per helical turn. The negative topoisomers observed *in vitro* at low protein:DNA ratios therefore probably result from this interaction of HMf with DNA in the absence of toroidal wrapping. The positive

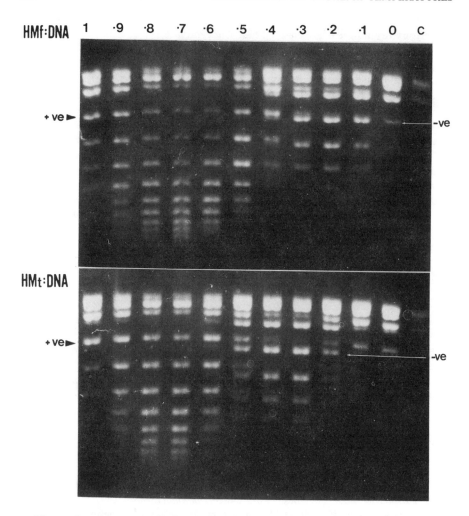

Figure 1. Agarose gel electrophoretic separations of pUC19 topoisomers. Relaxed, closed circular pUC19 DNA molecules (track C) were mixed, at the protein:DNA mass ratios indicated above each track, with HMf or HMt isolated for *Methanothermus fervidus* and *Methanobacterium thermoautotrophicum* ΔH, respectively. Following topoisomerase I treatment, the DNA molecules were deproteinized and topoisomers separated by electrophoresis through 1.5% (w/v) agarose gels run at 1.5v/cm in 90 mM Tris-borate/2.5 mM EDTA, pH8.3 for 16 hr. at room temperature. The majority of topoisomers in tracks 1 through 0.5 were positively supercoiled whereas in tracks 0.4 to 0.2 a mixture of positively and negatively supercoiled topoisomers were present. These configurations were confirmed by two dimensional agarose gel electrophoresis using ethidium bromide to amplify the separation of positive and negative topoisomers in the second dimension. (-ve, negative topoisomers; +ve, positive topoisomers)

topoisomers observed at higher protein to DNA mass ratios, at which HMf molecules already bound to the DNA interact, probably do result from the DNA being wrapped in a positive toroidal supercoil around a core of HMf. A detailed model for HMf/DNA complex formation, based on these arguments, has been presented (88).

To determine if positive toroidal supercoiling by DNA binding proteins correlated with the presence of reverse gyrase, a comparison of toroidal supercoiling by HMf from *M. fervidus* (contains reverse gyrase) and HMt, a similar DNA binding protein purified from *M. thermoautotrophicum* ΔH (lacks reverse gyrase) was carried out. HMf and HMt were both purified as previously described for HMf (*44,48*) and incubated over a range of protein to DNA mass ratios with relaxed, covalently closed, circular molecules of pUC19 DNA. Plectonemic supercoils which formed to compensate for toroidal wrapping were removed by wheat germ topoisomerase I and the DNA molecules then deproteinized. The resulting topoisomers were separated by electrophoresis through agarose gels in the absence of ethidium bromide. Very similar patterns of topoisomers were produced by HMf and HMt binding to pUC19 (Figure 1). The production of predominantly positively supercoiled topoisomers at mass ratios above 0.3:1 for HMf and above 0.4:1 for HMt and negatively supercoiled topoisomers below these mass ratios was confirmed by two dimensional agarose gel electrophoresis. DNA binding proteins that can wrap DNA in positive toroidal supercoils are therefore not found exclusively in hyperthermophilic organisms containing reverse gyrase. Positive toroidal wrapping might have evolved in these two thermophilic methanogens to balance the effects that their high intracellular salt concentrations (*78*) must have on the stability of their genomes. Positive toroidal wrapping of some genomic regions around HMf or HMt should result in a compensating increase in negative plectonemic superhelicity in the remaining protein-free regions of the genome. This increased negative superhelicity may allow the DNA strands in these protein-free regions to separate more easily in the presence of a high intracellular salt concentration.

To relate the HMf-DNA interactions observed *in vitro* to the situation *in vivo* it was necessary to know the intracellular ratio of HMf to DNA and the percentage of HMf molecules which actually bind to DNA in a mixture of HMf and DNA molecules. As free HMf is soluble in 70% ethanol but HMf/DNA complexes are precipitated by 70% ethanol, a co-precipitation assay for HMf binding to DNA was developed. HMf and DNA were mixed and, after ethanol addition and centrifugation, the amounts of HMf in the pellet and remaining in the supernatant were measured using anti-HMf antibodies in an ELISA assay. Approximately 90% of the input HMf molecules bound to the DNA using the binding conditions routinely employed (*44,48*). Experiments are currently in progress to determine if this remains the case at higher temperatures and at salt concentrations which mimic the conditions inside *M. fervidus* cells. The anti-HMf antibodies have also been used in a quantitative Western blot analysis to determine precisely the intracellular concentration of HMf. Cells from exponentially growing cultures of *M. fervidus* were harvested at an A_{580} of 0.5, lysed by boiling in 4% SDS and the resulting extract treated with DNaseI. Proteins in this extract were separated by electrophoresis through an SDS-polyacrylamide gel and then transferred to a nitrocellulose filter which was incubated with the anti-HMf antibodies. A standard curve for HMf quantitation was obtained using purified HMf and scanning densitometry of the resulting Western blot. The results obtained with the total cell extract demonstrated that the concentration of HMf *in vivo* is between 10,000 and 20,000 molecules per 1.5 Mbp genome, corresponding to an HMf to DNA mass ratio of ~0.3:1. This is very close to the value of 0.25:1, determined previously for this ratio, based on the amount of HMf and DNA obtained from *M. fervidus* cells by purification (*44,48*) and predicts that HMf-directed positive toroidal wrapping of genomic DNA could occur *in vivo*.

Further Analysis of the HMf:DNA Interaction

A variety of molecular probes, including the hydroxyl radical generated by the Fenton reaction using Fe (II)EDTA, have been developed to examine the structure of DNA in sequence-neutral reactions (*80-83*). Investigations of HMf/DNA complexes using the hydroxy radical footprinting procedure have demonstrated that in these complexes the DNA molecule is significantly protected from hydroxyl radical cleavage and has a helical periodicity of ~11bp at all protein to DNA mass ratios investigated (88).

The molecular architecture of the HMf/DNA complexes has now also been characterized using DNase I as the investigative probe. Complexes were formed between HMf and ^{32}P-end-labeled, linear pUC19 DNA molecules, at mass ratios ranging from 0.2:1 to 1:1, and subjected to digestion with DNase I (*11,83*). Because of the nuclease protection afforded the DNA by HMf binding, it was necessary to use 3x and 10x more DNase I to obtain footprints in the presence of HMf, at mass ratios of 0.2:1 and 1:1, as compared to the amount of DNaseI needed in control digestions in the absence of HMf. The results of a DNAse I footprinting experiment are shown in Figure 2. The patterns of DNA fragments generated at HMf to DNA mass ratios of 1:1 (lane A) and 0.2:1 (lane B) are similar and significantly different from the pattern generated in the absence of HMf (lane C). Longer digestion times, at all HMf:DNA ratios, do however produce fragment patterns which resemble those of the control. This was expected as HMf/DNA complexes are known to be dynamic and are continually undergoing association and disassociation (*48*). All sites should become available for DNase I cleavage over a period of time. As DNase I cleavage has some sequence specificity the fragment patterns obtained in the presence of HMf using DNaseI have only barely discernable helical periodicities.

HU Complementation

To test for functional relatedness between HMf and HU, the cloned *hmf*B gene has been expressed in an *E. coli* strain deleted for the two genes, *hup*A and *hup*B, that encode the two subunits of HU. Plaque formation by Mu phage is reduced by

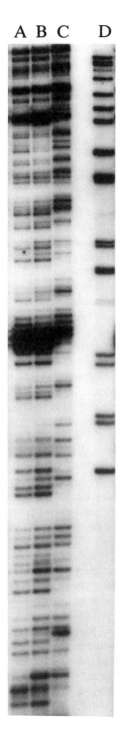

A B C   D

Figure 2. DNaseI footprinting of HMf binding to pUC19 DNA. Plasmid DNA was linearized by digestion with *EcoR*1 and [^{32}P]-end-labeled using Sequenase and α[^{32}P]-dATP. The DNA was incubated with HMf at HMf:DNA mass ratios of 1:1 (track A) and 0.2:1 (track B). Digestion products of control DNA without HMf are shown in track C. DNaseI digestion was initiated by the addition of 2mM CaCl$_2$, and stopped by addition of 100mM EDTA and phenol. Autoradiograms obtained following DNaseI footprinting, acrylamide gel electrophoresis and autoradiography were quantitated by scanning densitometry. ^{32}[P]-end-labeled *Msp*I fragments of pBR322 DNA were used as size standards (track D).

seven orders of magnitude on this double mutant (*84*), however *hmf*B expression did not increase the ability of Mu to productively infect this strain. HMf suppression of the other phenotypes of the *hup*A, *hup*B double mutant (*25,26,84*) is currently being evaluated.

Conclusions

The overall superhelicity of the *M. fervidus* genome *in vivo* must be determined by HMf-directed nucleosome-like wrapping, DNA binding by sequence specific proteins, the activity of topoisomerases, temperature and the intracellular concentration of solutes, most notably the concentration of K_3cDPG. The HMf to DNA ratio *in vivo* of 0.3:1 is sufficient to wrap at least one third of the *M. fervidus* genome in positive toroidal supercoils and seems likely to play a major role in the architecture of the genome and in providing significant thermal protection. Genome topology must be dynamic and superhelicity represents energy that can be used to regulate protein/DNA interactions and structural conformations of the DNA (*85,86*). HMf binding therefore almost certainly also plays a role indirectly if not directly in gene regulation, DNA replication and recombination. Developing *in vitro* systems employing physiologically relevant conditions to investigate these activities is now the real challenge.

The structure of the HMf/DNA complex which forms *in vitro* depends on the protein to DNA mass ratio rather than the concentration of these reactants. It is apparently the number of protein molecules that bind per DNA molecule that determines if toroidal wrapping occurs. HMf molecules must first associate with DNA and then protein/protein interactions must occur to direct the positive toroidal supercoiling. A tetrameric core of HMf molecules has been proposed in the model for the HMf/DNA complex and this does not form in the absence of DNA (*88*). Positive toroidal wrapping of the *M. fervidus* genomic DNA may function to counteract the tendency towards positive plectonemic supercoiling of this molecule caused by the high intracellular ionic environment and by reverse gyrase activity. As binding of HMf molecules increases the helical periodicity of the DNA to 11.0 bp, this binding should be favored thermodynamically on positively supercoiled molecules. It is tempting therefore to speculate that unconstrained regions of the *M. fervidus* genome should be positively supercoiled. To accommodate the negative toroidal wrapping of DNA around reverse gyrase and its preference for ssDNA, features which indicate that this enzyme interacts preferentially with negatively supercoiled DNA, we would suggest by analogy with the role proposed for DNA gyrase by Liu and Wang (*38*) that *in vivo* reverse gyrase might recognize and remove negative supercoils rather than introduce positive supercoils.

The tight bending of the DNA molecule needed for nucleosome assembly has led to the idea that major determinants of nucleosome positioning may be the inherent curvature and flexibility of DNA sequences. In the nucleosome G/C rich sequences occur predominantly where the minor groove is oriented away from the direction of DNA curvature and A/T sequences occur more frequently where the minor groove faces inwards (*32,36,53,85,86*). The possibility that the locations at which HMf/DNA complexes form might also be determined by sequence dependent DNA curvatures remains to be investigated. It is, however, well established and intriguing that intergenic regions in *M. fervidus* are very A/T rich and many contain presumably 'bent' oligo-A sequences *(41,87)*. When sufficient DNA sequences are available from *M. fervidus* and *M. thermoautotrophicum*, their detailed examination for inherent bending periodicities should indicate if these genomic DNAs are organized to facilitate positive toroidal wrapping *in vivo* by HMf and HMt.

Acknowledgements

The authors' research reported here, undertaken at The Ohio State University, was supported by contract N00014-86-K-0211 from the U.S. Office of Naval Research.

Literature Cited

1. Cairns, J. *J. Mol. Biol.* **1963**, *6*,208-213.
2. Bachmann, B.J. *Microbiol. Rev.* **1983**, *47*,180-230.
3. Smith, C.L.; Econome, J.G.; Schutt, A.; Kico, S.; Cantor, C.R. *Science* **1987**, *236*,1448-1453.
4. Kohara, Y.; Akiyama, K.; Isono, K. *Cell* **1987**, *50*,495-508.
5. Kellenberger, E. In: *The Bacterial Chromosome*; Drlica K.; Riley, M., Eds.; American Society for Microbiology, Washington, D.C. 20005 1990, pp. 173-186.
6. Bohrmann, B.; Villiger, W.; Johansen, R.; Kellenberger, E. *J. Bacteriol.* **1991**, *173*, 3149-3158.
7. Delius, H.; Worcel, A. *J. Mol. Biol.* **1974**, *82*, 107-109.
8. Kavenoff, R.; Ryder, O. *Chromosoma* **1976**, *59*,13-25.
9. Lyderson, K.; Pettijohn, D.E. *Chromosoma* **1977**, *62*,199-215.
10. Pettijohn, D.E.; *J. Biol. Chem.* **1988**, *263*, 12793-12796.
11. Broyles, S.S.; Pettijohn, D.E. *J. Mol. Biol.* **1986**, *187*, 47-60.
12. Drlica, K.; Rouvière-Yaniv, J. *Microbiol. Rev.* **1987**, *51*, 301-309.
13. Dixon, N.; Kornberg, A. *Proc. Natl. Acad. Sci. USA* **1984**, *81*,424-428.
14. Flashner, Y.; Gralla, J. *Cell* **1988** *54*, 713-721.
15. Hodges-Garcia, Y.; Hagerman, P.; Pettijohn, D.E. *J. Biol. Chem.* **1989**, *264*, 14621-14623.
16. Skarstad, K.; Baker, T.A.; Kornberg, A. *The EMBO Journal* **1990**, *9*, 2341-2348.
17. Varshavsky, A.; Nedospasov, S.A.; Bakayev, V.V.; Bakayeva, T.G.; Georgiev, G.P. *Nucleic Acids Research* **1977**, *4*, 2725-2745.
18. Pettijohn, D.E.; Hodges-Garcia, Y. In: *The Bacterial Chromosome*; Drlica, K.; Riley, M., Eds. American Society for Microbiology, Washington D.C. 20005. 1990, pp. 241-245.
19. Bliska, J.; Cozzarelli, N. *J. Mol. Biol.* **1987**, *194*, 205-218.

20. Pettijohn, D.E.; Pfenninger, O. *Proc. Natl. Acad. Sci.* USA **1980**, *77*, 1331-1335.
21. Yang, Y.; Ames, G.F-L. *Proc. Natl. Acad. Sci.* USA **1988**, *85*, 8850-8854.
22. Uemura, T.; Ohkura, H.; Adachi, Y.; Morino, K.; Shizaki, K.; Yanagida, M. *Cell* **1987** *50*, 917-925.
23. Wada, M.; Kano, Y.; Ogawa, T.; Okazaki, T.; Imamoto, F. *J. Mol. Biol.* **1988**, *204*, 581-591.
24. Schmid, M.B. *Cell* **1990**, *63*, 451-453.
25. Kano, Y.; Goshima, N.; Wada, M.; Imamoto, F. *Gene* **1989**, *76*, 353-358.
26. Ogura, T.; Niki, H.; Kano, Y.; Imamoto, F.; Hiraga, S. *Molec. Gen. Genet.* **1990**, *220*, 197-203.
27. Durrenberger, M.; Bjornsti, M.A.; Uetz, T.; Hobot, J.A.; Kellenberger, E. *J. Bacteriol.* **1988**, *170*, 4757-4768.
28. Shellman, V.L.; Pettijohn D.E. *J. Bacteriol.* **1991**, *173*, 3047-3054.
29. Kornberg, R. *Annu. Rev. Biochem.* **1977**, *46*, 931-954.
30. Felsenfeld, G. *Nature* (London) **1978**, *271*, 1115-122.
31. Drew, H.R.; Travers, A.A. *J. Mol. Biol.* **1985**, *186*, 773-790.
32. Travers, A.A.; Klug, A. *DNA Topology and Its Biological Effects.* Cold Spring Harbor Laboratory Press, Cold Spring Harbor, NY, **1990**; pp. 57-106.
33. Rhodes, D.; Klug, A. *Nature* **1980**, *286*, 573-578.
34. Tullius, T.D.; Dombroski, B.A. *Science* **1985**, *230*, 679-681.
35. Hayes, J.J.; Tullius, T.D.; Wolffe, A.P. *Proc. Natl. Acad. Sci.* USA **1990**, *87*, 7405-7409.
36. Hayes, J.J.; Clark, D.J.; Wolffe, A.P. *Proc. Natl. Acad. Sci.* USA **1991**, *88*, 6829-6833.
37. Klug, A.; Lutter L.C. *Nucleic Acids Res.* **1981**, *9*, 4267-4283.
38. Liu, L.F.; Wang, J.C. *Proc. Natl. Acad. Sci* USA **1987**, *84*, 7024-7027.
39. Clark, D.J.; Felsenfeld, G. *The EMBO J.* **1991**, *10*, 387-395.
40. Churchill, M.E.A.; Travers, A.A. *TIBS* **1991**, *16*, 92-97.
41. Brown, J.W.; Daniels, C.J.; Reeve, J.N. *CRC Crit. Rev. in Microbiol.* **1989**, *16*, 238-287.
42. Searcy, D.G.; Stein, D.B. *Biochim. Biophys. Acta* **1980**, *609*, 180-195.
43. Dijk, J.; Reinhardt, R. In: *Bacterial Chromatin*; Gualerzi, C.O.; Pon, C.L., Eds. Springer Verlag, Berlin **1986**, pp 186-219.
44. Sandman, K.M.; Krzycki, J.A.; Dobrinski, B.; Lurz, R.; Reeve, J.N. *Proc. Natl. Acad. Sci.* **1990**, *87*, 5788-5791.
45. DeLange, R.J.; Green, G.R.; Searcy, D.G. *J. Biol. Chem.* **1981**, *256*, 900-904.
46. Laine, B.; Culard, F.; Maurizot, J-C.; Sautière, P. *Nucleic Acids Res.* **1991**, *19*, 3041-3045.
47. DeLange, R.J.; Williams, L.C.; Searcy, D.G. *J. Biol. Chem.* **1981**, *256*, 905-911.
48. Krzycki, J.A.; Sandman, K.M.; Reeve, J.N. In: *Proceedings of the 6th International Symposium on the Genetics of Industrial Microorganisms*;

Heslot, H.; Davies, J.; Florent, J.; Bobichon, L.; Durand, G; Penasse, L., Eds.; Société Française de Microbiologie, Strasbourg, France, **1990**; vol. 2; pp. 603-610.
49. Wang, J.C. *J. Mol. Biol.* **1971**, *55*, 523-533.
50. Wang, J.C. *J. Biol. Chem.* **1991**, *2266*, 6659-6662.
51. Wang, J.C. *Annu. Rev. Biochem.* **1985**, *54*, 665-697.
52. Drlica, K. *Microbiol. Rev.* **1984**, *48*, 273-289.
53. Gellert, M. *Annu. Rev. Biochem.* **1981**, *50*, 879-910.
54. Maxwell, A.; Gillert, M. *Adv. Protein Chem.* **1986**, *38*, 69-107.
55. Liu, L.F.; Wang, J.C. *Cell* **1978**, *15*, 979-984.
56. Reece, R.J.; Maxwell, A. *Nucleic Acids Res.* **1991**, *19*, 1399-1405.
57. Kirkegaard, K.; Pflugenfelder, G.; Wang, J.C. Cold Spring Harbor Symp. *Quant. Biol.* **1984**, *49*, 411-419.
58. Kirkegaard, K.; Wang, J.C. *J. Mol. Biol.* **1985**, *185*, 625-637.
59. Depew, R.E.; Wang, J.C. *Proc. Natl. Acad. Sci.* USA **1975**, *72*, 4275-4279.
60. Pullybank, D.E.; Shure, M.; Tang, D.; Vinograd, J.; Vosberg, H-P. *Proc. Natl. Acad. Sci.* USA **1975**, *72*, 4280-4284.
61. Hsieh, L-S.; Rouvière-Yaniv, J.; Drlica, K. *J. Bacteriol.* **1991**, *173*, 3914-3917.
62. Kikuchi, A.; Asai, K. *Nature* **1984**, *309*, 677-681.
63. Slesarev, A.A.; Kozyavkin, S.A. *J. Biomol. Structure and Dynamics* **1990**, *7*, 935-942.
64. de la Tour, C.B.; Portemer, C.; Nadal, M.; Stetter, D.O.; Forterre, P.; Duguet, M. *J. Bacteriol.* **1990**, *172*, 6803-6808.
65. Bouthier de la Tour, C.; Portemer, C.; Huber, R.; Forterre, P.; Dugnet, M. *J. Bacteriol.* **1991**, *173*, 3921-3923.
66. Collin, R.G.; Morgan, H.W.; Musgrave, D.R.; Daniel, R.M. *FEMS Microbiol. Lett.* **1988**, *55*, 235-240.
67. Nakasu, S; Kikuchi, A. *EMBO J.* **1985**, *4*, 2705-2710.
68. Collin, R.G. Ph.D. Thesis **1990**, The University of Waikato, Hamilton, New Zealand.
69. Slesarev, A.A., *Eur. J. Biochem.* **1988**, *173*, 395-399.
70. Kovalsky, O.I.; Kozyavkin, S.A.; Slesarev, A.A. *Nucleic Acis Res.* **1990**, *118*, 2801-2806.
71. Jaxel, C.; Nadal, M.; Mirambeau, G.; Forterne, P.; Takahashi, M.; Duguct, M. *EMBO J.* **1989**, *8*, 3135-3139.
72. Kikuchi, A. In: *DNA Topology and its Biological Effects*; Cozzarelli, N.R.; Wang, J.C., Eds. Cold Spring Harbor Laboratory Press. Cold Spring Harbor NY, **1990** pp. 285-298.
73. Krogh, S.; Mortensen, U.H.; Westergaard, O.; Bonven, B.J. *Nucleic Acids Research* **1991**, *19*, 1235-1241.
74. Nadal, M.; Mirambeau, G.; Forterre, P.; Reiter, W-D.; Duguet, M. *Nature* **1986**, *321*, 256-258.
75. Murray, P.J. M.Sc. Thesis **1988**, University of Waikato, Hamilton, New Zealand.

76. Stetter, K.O.; Thomm, M.; Winter, J.; Wilgruber, G.; Hüber, H.; Zillig, W.; Janekovic, D.; König, H.; Palm, P.; Wunderl, S. *Zentralbl. Bakteriol. Mikrobiol.* Hyg. I. Abt. Orig. **1981**, *C2*, 166-178.
77. Zillig, W.I.; Holz, I.; Klenk, H-P.; Trent, J.; Wunderl, S.; Janekovic, D.; Imsel. E.; Haas, B. *Syst. Appl. Microbiol.* **1987**, *9*, 62-70
78. Hensel, R.; König, H. *FEMS Microbiology Letters* **1987**, *49*, 75-79.
79. Woese, C.R.; Olsen, G.J. *System Appl. Microbiol.* **1986**, *7*, 161-177.
80. Jessee, B.; Garguilo, G.; Razvi, F.; Worcel, A. *Nucl. Acids Res.* **1982**, *10*, 5823-5834.
81. Ward, B.; Skorobogaty, A.; Dabrowiak, J.C. *Biochemistry* **1986**, *25*, 6875-6883.
82. Uchida, K.; Pyle, A.M.; Morii, T.; Barton, J.K. *Nucl. Acids Res.* **1989**, *17*, 102259-10278.
83. Tulius, T.D.; Dombroski, B.A.; Churchill, M.E.A.; Kam, L. *Methods Enzymol.* **1987**, *155*, 537-558.
84. Imamoto, F; Kano, Y. In *The Bacterial Chromosome*; Drlika, K.; Riley, M., Eds.; American Society for Microbiology Washington D.C. **1990** pp.241-245.
85. Travers, A.A. In *Nucleic Acids and Molecular Biology.* Eckstein, F.; Lilley, D.M.J., Eds.; Springer Verlag. Berlin **1988** Vol. 2, pp 136-148.
86. Travers, A.A. *Cell* **1990**, *60*, 177-180.
87. Haas, E.S.; Brown, J.W.; Daniels, C.J.; Reeve, J.N. *Gene* **1990**, *90*, 51-59.
88. Musgrave, D.R.; Sandman, K.M.; Reeve, J.N. *Proc. Natl. Acad. Sci. USA* **1991**, *88*, 10397-10401.

RECEIVED January 15, 1992

Chapter 13

Applications of Thermostable DNA Polymerases in Molecular Biology

E. J. Mathur

Research and Development Division of Stratagene Inc., 11099 North Torrey Pines Road, La Jolla, CA 92037

With the advent of the polymerase chain reaction (PCR), thermostable DNA polymerases have begun to play an important role in molecular biology. We have recently isolated and cloned a novel DNA polymerase from the hyperthermophilic marine archaebacterium, *Pyrococcus furiosus* (*Pfu*). The multifunctional enzyme has been shown to function effectively in PCR. Moreover, results from fidelity studies indicate that amplification reactions performed with *Pfu* DNA polymerase result in PCR products with 12 fold less mutations than products from similar amplifications performed with *Thermus aquaticus* (*Taq*) DNA polymerase. These finding may have important ramifications for PCR-based procedures which require high fidelity DNA synthesis.

Technological advancements are made in one of two ways: evolution or revolution. New methodologies can be developed through improvement and refinement of existing technology or can result from a revolutionary discovery. One such revolution was the discovery of the polymerase chain reaction (*1*). The polymerase chain reaction (PCR) is a powerful biochemical technique which can selectively amplify a specific DNA sequence out of complex mixtures of nucleic acids. PCR has revolutionized the way in which molecular biologists manipulate nucleic acids.

While driving along the coastline in northern California during the spring of 1983, Kerry Mullis first conceived the concept of the polymerase chain reaction. It was not until five years later, that the use of a thermostable DNA polymerase in PCR was first described (*2*). The PCR technique, also known as *in vitro* gene amplification, was greatly improved by incorporation of a

0097–6156/92/0498–0189$06.00/0

thermostable DNA polymerase into the amplification protocol. The ability of a DNA polymerase enzyme to remain active following repeated exposures to temperatures approaching 100°C allowed the once tedious PCR process to become automated and amenable to revolutionizing existing molecular biology techniques. PCR has facilitated the development of gene characterization and molecular cloning technologies including the direct sequencing of PCR amplified DNA, the determination of allelic variation and the detection of infectious and genetic disease disorders.

PCR also served to introduce molecular biologists to the amazing potential of thermostable enzymes. During the past several years, the biotechnology community has become actively engaged in the development of new procedures and the modification of existing techniques which utilize the thermostable counterparts of the mesophilic enzymes currently used in molecular biology today. Standard recombinant DNA methodologies which have been traditionally inefficient or time-consuming have the potential of being improved or replaced with techniques which exploit enzymes that function and remain active at elevated temperatures.

This chapter will review PCR theory and the enzymatic properties of thermostable DNA polymerases which make them well suited for use in the polymerase chain reaction. The discussion will include the discovery and applications of a remarkable thermostable DNA polymerase isolated from a hyperthermophilic archaebacterium which has been shown to replicate DNA *in vitro* with greater than ten fold higher accuracy than the alternative enzymes currently used in PCR today.

The chapter will conclude with a synopsis of the emerging trends and the enzymatic requirements for second generation PCR technologies, together with a discussion of other important thermostable enzymes and their potential applications in the field of molecular biology. This section will also include a brief introduction to the Human Genome Initiative and the potential contributions that thermostable DNA polymerases can make to DNA sequencing technologies.

DNA Polymerases and the Polymerase Chain Reaction

In order to grasp the concept and theories encompassed by the PCR technique, it is important to have a fundamental understanding of the mechanism of DNA polymerases. DNA polymerases are a class of enzymes which catalyze the stepwise addition of deoxyribonucleoside 5' triphosphates to the 3'-OH terminus of a polynucleotide chain complementarily base paired to a second, template strand (Figure 1). The reaction catalyzed by DNA polymerases result in the formation of a phosphodiester bond between the 3'-OH group of the terminal nucleotide with the 5'-phosphate of the incoming deoxyribonucleoside triphosphate (3). The nascent strand is thus synthesized in the 5' to 3' direction. The order in which the deoxyribonucleotides are added to the growing strand is dictated by base-pairing to the template strand. DNA polymerases require both primer and template DNA in addition to all four nucleoside triphosphates and magnesium ions in order to catalyze the polymerase reaction.

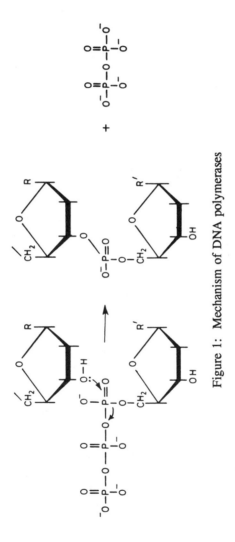

Figure 1: Mechanism of DNA polymerases

At the nucleotide insertion level, accurate DNA synthesis occurs as a result of the ability of the DNA polymerase enzyme to discriminate between complementary and non-complementary incoming nucleotides prior to the formation of a phosphodiester bond (4). An additional mechanism whereby certain DNA polymerases can achieve high fidelity DNA synthesis involves a property known as exonucleolytic proofreading activity. In addition to 5' to 3' DNA polymerase activity, some DNA polymerases also possess a 3' to 5' exonuclease dependent proofreading activity . These multifunctional DNA polymerases typically replicate DNA with substantially higher fidelity than DNA polymerases lacking 3' to 5' exonuclease activity (5-10). Multifunctional DNA polymerases possessing 3' to 5' exonuclease activity will preferentially excise misincorporated nucleotides and permit the incorporation of the correct, complementary nucleotide prior to chain elongation. Thus, polymerases which exhibit proofreading activity replicate DNA with high fidelity by at least two mechanisms: 1) discrimination at the nucleotide incorporation step; also known as base insertion fidelity and 2) proofreading following misincorporation of a non-complementary nucleotide (11). DNA polymerase replication fidelity is the key factor to consider when screening new thermostable DNA polymerases for high fidelity PCR applications.

The polymerase chain reaction is performed by repeated cycles of heat denaturation of a DNA template containing the target sequence, annealing of opposing primers to the complimentary DNA strands with the 3'-hydroxyl ends facing each other, and extension of the annealed primers with a DNA polymerase (Figure 2). Multiple PCR cycles result in the exponential amplification of the nucleotide sequence delineated by the flanking amplification primers; the PCR primers are incorporated into the ends of all the amplification products (Figure 3).

An important modification of the original protocol was the substitution of thermostable *Thermus aquaticus (Taq)* DNA polymerase in place of the Klenow fragment of *E. coli* DNA polymerase I (2). The incorporation of a thermostable DNA polymerase into the PCR protocol obviates the need for repeated enzyme additions as the Klenow fragment of *E. coli* polymerase I is irreversibly denatured at the temperatures required for DNA template denaturation during PCR. *Taq* DNA polymerase, however, maintains roughly 75% of it's DNA polymerase activity following a 1 hr. incubation at 95°C (unpublished data). In addition, because *Taq* DNA polymerase has a optimum temperature for activity near 75°C, the PCR cycling conditions have been altered to include elevated annealing and extension temperatures. Increasing the cycling temperatures during PCR serves to enhance the specificity of primer:template associations and results in an increase in yield of the desired PCR products. *Taq* polymerase thus increases the specificity, simplicity and overall vigor of the polymerase chain reaction which in turn has led to the automation of PCR.

Although *Taq* polymerase was the first thermostable polymerase utilized in PCR and still is used in the vast majority of the PCR performed today, it has a fundamental drawback; purified *Taq* polymerase enzyme does not possess 3' to 5' exonuclease dependent proofreading activity and therefore PCR amplification

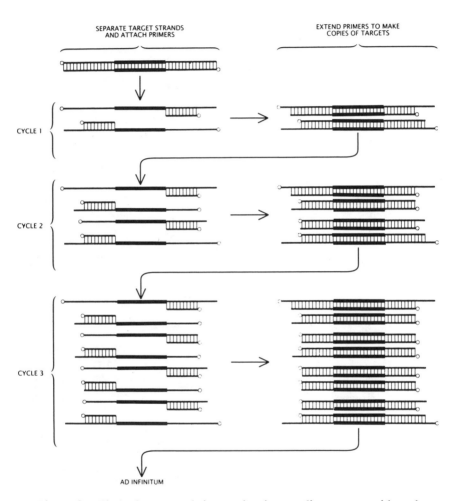

Figure 2: The polymerase chain reaction is a cyclic process; with each cycle, the number of target molecules doubles. The strands in each targeted DNA duplex are separated by heating and then cooled to allow primers to bind to them. Next, DNA polymerases extend the primers by adding nucleotides to them. In this way, duplicates of the original DNA-strand targets are produced. (Reproduced with permission from ref. *33*. Copyright 1990 Scientific American.)

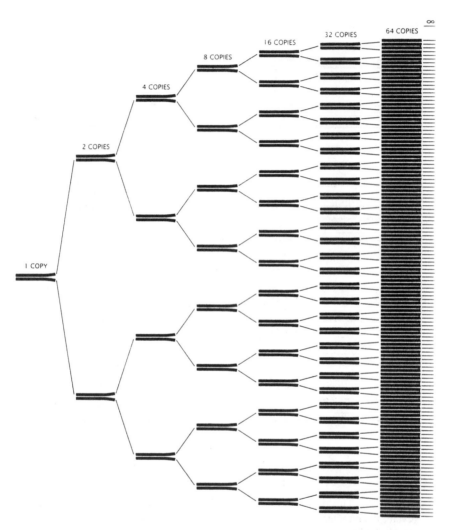

Figure 3: The polymerase chain reaction is a simple technique for copying a piece of DNA in the laboratory with readily available reagents. Because the number of copies increases exponentially, more than 100 billion can be made in only a few hours. (Reproduced with permission from ref. *33*. Copyright 1990 Scientific American.)

products are synthesized with relatively low fidelity (*12*). The observed error rate deduced for *Taq* DNA polymerase is indeed high; estimates range from 2 x 10^{-4} during PCR (*2,13*) to 2 x 10^{-5} for base substitution errors produced during a single round of DNA synthesis of the *lacZ* gene (*14*). Polymerase induced mutations acquired during PCR increase arithmetically as a function of cycle number. For example, if an average of two mutations occur during one cycle, 20 mutations will occur after 10 cycles and 40 will occur after 20 cycles. Each mutant and wild type template DNA molecule will be amplified exponentially during PCR and thus a large percentage of the resulting amplified products will harbor mutations. For many PCR applications, high fidelity amplification is not critical. Polymerase induced mutations incurred during the amplification process result in a population of amplified molecules in which each nucleotide within the amplified sequence is correct in the vast majority of the molecules. Thus, in direct sequencing of PCR amplified products or nucleic acid hybridization procedures, mutations within the sequence will generally not be detected. However, mutations introduced by *Taq* polymerase during amplification have hindered PCR applications which require high fidelity DNA synthesis. These applications include the direct cloning of PCR amplified products (*15*), PCR-based procedures for high efficiency double-stranded mutagenesis (*16*), amplification techniques designed to detect specific point mutations (*17*), the study of allelic polymorphisms of individual mRNA transcripts (*18,19*) and the characterization of the allelic states of single sperm cells (*20*) or single DNA molecules (*21,22*).

Pyrococcus furiosus (*Pfu*) DNA Polymerase

We have recently isolated and characterized a novel thermostable DNA polymerase from the hyperthermophilic marine archaebacterium, *Pyrococcus furiosus* (*23-25*). This monomeric, multifunctional enzyme has a molecular weight of 90,113 and possesses both DNA polymerase and 3' to 5' exonuclease dependent proofreading activities. The polymerase is extremely thermostable with a temperature optimum near 75°C and it retains greater than 95% of its activity after a 1 hr. incubation at 95°C. The purified enzyme functions effectively in the polymerase chain reaction. Moreover, results from our fidelity studies indicate that PCR performed with the *Pfu* DNA polymerase yields amplification products containing less than 10% of the number of mutations obtained from similar amplifications performed with *Taq* DNA polymerase. The remainder of this discussion will describe in more detail the enzymatic properties of *Pfu* DNA polymerase which make this remarkable enzyme well suited for high fidelity gene amplification techniques.

The thermostable DNA polymerase isolated from *P. furiosus* was purified to greater than 99% homogeneity as assessed by analysis of silver stained denaturing polyacrylamide gel electrophoresis (Figure 4). As shown in Table I, purified *Pfu* DNA polymerase possesses substantial 3' to 5' exonuclease activity; in contrast, *Taq* DNA polymerase exhibits no detectable 3' to 5' exonuclease activity. Very recently, we have succeeded in cloning the gene for *Pfu* DNA

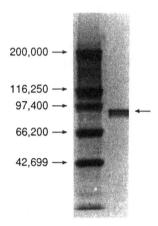

Figure 4: Denaturing polyacrylamide gel electrophoresis of *P. furiosus* DNA polymerase.

Table I: Polymerase and 3' to 5' exonuclease specific activities for purified *P. furiosus* and *T. aquaticus* DNA polymerases. (Reprinted with permission from ref. *23* . Copyright 1991 Elsevier Science Publishers BV.)

source of DNA polymerase	protein concentration (mgs/ml)	polymerase activity (units/mg)	3' to 5' exonuclease activity (units/mg)
P. furiosus	0.13 (±0.004)	31,713 (±1015)	9200 (±294)
T. aquaticus	0.30 (±0.015)	16,667 (± 850)	negligible

polymerase and expressing it in *E. coli (26)*. The activity of the recombinant DNA polymerase enzyme is identical to that of the polymerase isolated from the natural source. Moreover, the high temperature, anaerobic, large scale fermentations required for the growth of *P. furiosus* are no longer necessary for production of *Pfu* DNA polymerase.

To analyze the 3' to 5' exonuclease activity at the molecular level, we designed an assay to elucidate the mechanism by which DNA polymerases respond to a mismatched 3' terminus. Three events can occur when a DNA polymerase interacts with a mismatched 3' primer terminus: no extension of the mismatched primer, extension with incorporation of the mismatched nucleotide(s) or extension following excision of the 3' mismatched nucleotide(s). To address this question, a 35 base synthetic template was constructed along with four 15 base oligonucleotide primers complementary to the template, but containing 0, 1 , 2 and 3 mismatched nucleotides at the 3' termini. The template was designed with an internal *Eco*RI site which coincides with the mismatched nucleotides at the 3' end of the 15mer primers. The 5' end-labeled primers were annealed to the 35mer template prior to incubation with all four nucleoside triphosphates, magnesium ions and either *Pfu* or *Taq* DNA polymerase. Following extension, half of the reaction mixture was digested with *Eco*RI. *Pfu* DNA polymerase extended all the primers tested, as demonstrated by the presence of the 35 base labeled product before restriction digestion. Moreover, *Pfu* DNA polymerase excised the mismatched nucleotides from the 3' ends of the primers and incorporated the correct complementary nucleotides thus restoring the *Eco*RI site as demonstrated by the digestion of the 35 base products. In contrast, *Taq* DNA polymerase was only capable of extending the wild type (perfectly complementary) primer to form the 35 base product susceptible to digestion with *Eco*RI, and was unable to extend any of the mismatched primers (Figure 5).

Pfu DNA polymerase was also evaluated for use in the polymerase chain reaction. Amplification reactions were performed with many different primer:template combinations using either *Pfu* or *Taq* polymerase. In most cases tested, PCR performed with *Pfu* DNA polymerase produced resulted in reaction products comparable to those obtained with *Taq* polymerase. Several examples of PCR products obtained from amplification reactions with *Pfu* DNA polymerase are shown in Figure 6.

As previously mentioned, numerous studies indicate that 3' to 5' exonuclease activity enhances the fidelity of DNA synthesis *(5-10)*. In order to confirm these finding with *Pfu* DNA polymerase, we devised an assay which would determine the relative accuracies of *Pfu* and *Taq* polymerase during PCR. The assay is based on a modification of an *in vivo* mutagenesis assay which employes the lac repressor gene *(lacI)* as a mutagenic target *(27)*. Transgeneic mouse genomic DNA containing a *lacI* transgene was isolated from brain tissue. The DNA was amplified with either *Pfu* or *Taq* DNA polymerase. The amplification products were digested at unique *Eco*RI sites introduced at the 5' end of the PCR primers, cloned phage λGT10, packaged and plated on an *E. coli* strain containing the α-complementating portion of the *lacZ* gene (Figure 7). A percentage of the mutations incurred within the *lacI* gene during the amplification

Flowchart

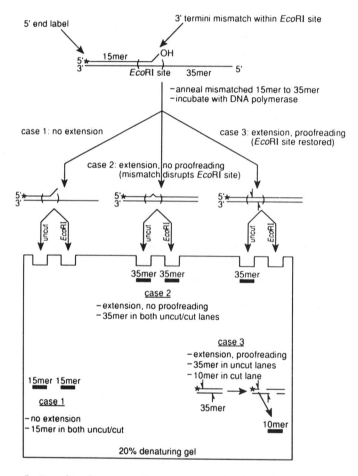

Figure 5: Proofreading assay diagram and results from the assay performed with *P. furiosus* and *T. aquaticus* DNA polymerases. See text for detailed description of experimental protocol. (Reproduced with permission from ref. *23*. Copyright 1991 Elsevier Science Publishers BV.)

Proofreading Assay

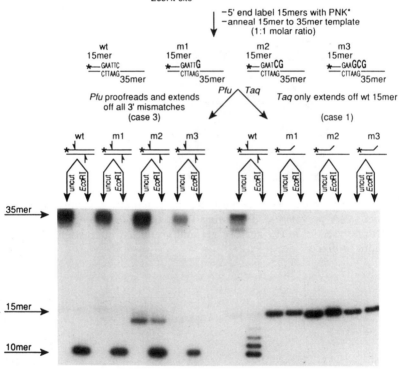

5' – TCCGGTCCC<u>GAATTC</u> –3' wt = 0 mismatch
5' – TCCGGTCCC<u>GAATTG</u> –3' m1 = 1 mismatch
5' – TCCGGTCCC<u>GAATCG</u> –3' m2 = 2 mismatch
5' – TCCGGTCCC<u>GAAGCG</u> –3' m3 = 3 mismatch
3' – AGGCCAGGG<u>CTTAAG</u>AGCAGCAGGCACGCGAACCG – 5' 35mer template
     *Eco*RI site

–5' end label 15mers with PNK*
–anneal 15mer to 35mer template
    (1:1 molar ratio)

wt   m1   m2   m3
15mer   15mer   15mer   15mer
*—GAATTC   *—GAATTG   *—GAATCG   *—GAAGCG
—CTTAAG35mer —CTTAAG35mer —CTTAAG35mer —CTTAAG35mer

Pfu proofreads and extends   *Pfu* *Taq*   *Taq* only extends off wt 15mer
off all 3' mismatches
(case 3)   (case 1)

wt   m1   m2   m3   wt   m1   m2   m3

35mer →

15mer →

10mer →

Figure 5. Continued

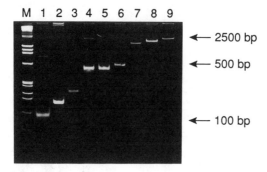

Figure 6: PCR products from amplification reactions performed with *P. furiosus* DNA polymerase. Following PCR, the reaction products were electrophoresed in a 1% agarose gel and visualized after staining with ethidium bromide.

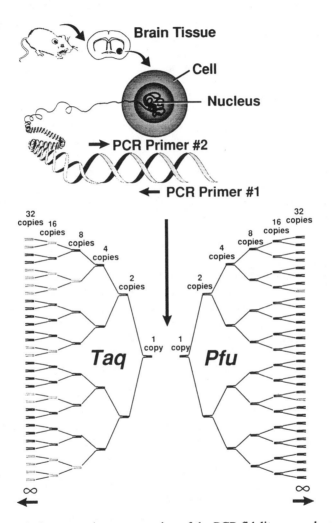

Figure 7: Diagrammatic representation of the PCR fidelity assay described in the text.

process result in a nonfunctional lac repressor protein. In the absence of active repressor protein, the α-complementing portion of β-galactosidase within the phage is expressed and can complement the Ω portion within the cell, generating a blue plaque phenotype when plated with top agar containing the chromogenic substrate, X-Gal. Therefore, the observed mutation frequency (*mf*; frequency of phenotypic mutants) can be calculated by dividing the number of blue plaques (lac*I*- mutants) by the total number of plaques scored (Table II).

$$\text{mutant frequency} = \text{blue plaques/total plaques scored}$$

The phage were also scored on plates with X-Gal and IPTG to demonstrate that 99% of the *Pfu* DNA polymerase amplified phage clones and 89% of the *Taq* DNA polymerase amplified phage clones were blue and therefore contained functional *lacZ* genes. Errors introduced during 10^5-fold amplification yielded an observed mutant frequency (*mf*) of 0.5% for *Pyrococcus* and 5.8% for *Taq* DNA polymerase. The background mutation frequency in *lacI* in mouse somatic tissues was determined to be $1.7 \pm 0.45 \times 10^{-5}$. Using these values, the error rate [mutations per nucleotide per cycle (ER)], was calculated for each DNA polymerase as:

$$ER = mf/bp \times d$$

where *bp* is the number of sites within the *lacI* gene known to yield a mutant phenotype and *d* is the number of effective duplications during PCR (16.6 for 10^5-fold amplification). The *bp* value of 182 was used because Gordon *et al* (29). have shown that 102 different sites within the *lacI* gene harbored various missense mutations after sequencing over 6000 spontaneous and induced *lacI*- mutants, and there are 80 nucleotide sites at which nonsense mutations can be produced. Our calculated error rate for *Taq* DNA polymerase is in close agreement with that reported by Eckert and Kunkle (*14*).

The average error rate was determined to be 1.6×10^{-6} for *Pfu* DNA polymerase and 2.0×10^{-5} for *Taq* polymerase. Thus, if a 1000 base pair sequence is amplified for 20 effective cycles with *Taq* polymerase, 40% of the amplification products will contain mutations. In contrast, if PCR is performed under identical conditions with *Pfu* DNA polymerase, only 3.2% of the PCR products will harbor mutations. These findings are significant for PCR techniques which require high fidelity DNA synthesis, including the direct cloning of PCR amplified products, PCR based procedures for high efficiency double stranded mutagenesis and amplification techniques designed to detect specific point mutations.

Future Applications for Thermostable Enzymes in Molecular Biology

While *Pfu* DNA polymerase will provide an obvious advantage over *Taq* DNA polymerase in PCR applications requiring high fidelity, there are other PCR-based technologies which require particular characteristics of thermostable DNA

TABLE II

Fidelity of *Pyrococcus furiosus* DNA polymerase and
Thermus aquaticus DNA polymerase in PCR fidelity assay

DNA polymerase	plaques scored total	mutant	mutant frequency (%)	error rate
Pfu				
PCR 1	9,044	62	0.60	2.0×10^{-6}
2	17,972	84	0.47	1.6×10^{-6}
3	15,903	56	0.35	1.2×10^{-6}
4	19,171	103	0.54	1.8×10^{-6}
Taq				
PCR 1	9,376	700	7.38	2.5×10^{-5}
2	10,190	538	5.20	1.7×10^{-5}
3	13,002	570	4.30	1.4×10^{-5}
4	14,640	916	6.25	2.1×10^{-5}

Table II: Tabulated results from the PCR fidelity assay described in the text. (Reprinted with permission from ref. *23* . Copyright 1991 Elsevier Science Publishers BV.)

polymerases which have not yet been described. One obvious need is the ability to PCR amplify very large fragments of DNA. The current size of standard PCR products ranges from 50 to 3000 base pairs with an upper limit of around 10 kilobases. Considerable research effort is currently being applied to screening for thermostable PCR polymerases capable of reproducibly amplifying templates greater than 10 kilobases.

The Human Genome project, a U.S. Government supported effort to sequence the entire human genome, has created a need for developing faster and more efficient methods for DNA sequencing. Using current technology, it has been estimated that project completion would require 200 technicians working full time for 20 years. Enormous research effort is being expended on the advancement of DNA sequencing technology which would facilitate the completion of the Human Genome project in a timely manner. One recent major advance in this field is known as cycle sequencing or linear amplification sequencing (29,30). Cycle sequencing involves the use of a thermostable DNA polymerase with the standard Sanger dideoxy chain terminator sequencing strategy. Cycle sequencing differs from standard sequencing in that with cycle sequencing, the extension reaction is cycled 30 times through template melting, primer annealing and primer extension steps, thus dramatically increasing the signal response. Unlike PCR, cycle sequencing results in the linear, not exponential amplification of the chain terminated, nested product molecules. The increased signal generated by cycle sequencing reactions allows low levels of unpurified DNA to be used as templates. For example, a single bacterial colony or bacteriophage plaque can be used directly in a cycle sequencing reaction with no purification. As in PCR, cycle sequencing also lends itself well to automation which will be a key element in the success of the Human Genome project. Currently, *Taq* DNA polymerase is the most widely used enzyme for cycle sequencing. However, due to *Taq*'s inherent limited processivity, the high K_m for the nucleotide analog incorporation and it's associated strand displacement activity, *Taq* DNA polymerase is not the ideal enzyme for cycle sequencing. Current efforts are underway to isolate new thermostable polymerases which have characteristics more ideal for cycle sequencing applications.

Thermostable DNA ligases have also become an important in molecular biology within the past few years. Recently several diagnostic techniques have been developed which require a thermostable DNA ligase capable of DNA ligase activity at elevated temperatures (30,31). The ligase chain reaction (LCR) was developed for the detection of mutations in genomic DNA. In the LCR technique, ligation of flanking primers is dependent on the accurate hybridization of the primers to complementary targets. If there are mutations which prevent correct hybridization, then the ligation event cannot occur. If the reaction is performed for repeated cycles with primers specific for both strands of the template DNA, the ligation products of each cycle serve as substrates for the next cycle of ligation. Thus, as in PCR, LCR results in the exponential amplification of the target template molecule. DNA ligase amplification techniques are rapidly becoming the method of choice for detection of single basepair substitutions in many DNA diagnostic applications.

There is also a need for thermostable reverse transcriptases; polymerases which will synthesize DNA from an RNA template. Traditionally, construction of genomic libraries is performed with mesophilic murine or avian retroviral reverse transcriptases. One major drawback of current methodologies is that the secondary structure elements present in RNA prevent certain sequences from being faithfully copied by the mesophilic reverse transcriptases. Thermostable reverse transcriptases, which function at elevated temperatures, would conceivably reduce or minimize the secondary structures present in single-stranded RNA molecules and thus allow the reverse transcriptase enzyme to freely translocate unimpaired through these regions of secondary structure. Thus, significant improvement in the quality of cDNA library construction could be accomplished by the incorporation of a thermostable reverse transcriptase into the protocol.

In summary, thermostable enzymes have become very important to molecular biologists of the 1990's. Of all the potential thermostable nucleic acid modifying enzymes, DNA polymerases are the class of thermophilic enzymes most exploited in biotechnology today. However, thermostable pyrophosphatases, alkaline phosphatases, reverse gyrases, DNA ligases, RNA polymerases and restriction endonucleases may all find important roles in biotechnology of the future.

Acknowledgments

The work described in this chapter would not have be possible without the experimental expertise of Dan Shoemaker, Kelly Lundberg, Kirk Nielson, and Mark Bergseid. I would like to also thank Mike Adams, Frank Robb and Joseph Sorge for their contributions towards a better understanding of thermostable enzymes.

Literature Cited:

1.Mullis,K.B.; Faloona, F.A. *Methods in Enzymology* **1987**, *155*:335-350.
2.Saiki,R.K.; Gelfand,D.H.; Stoffel,S.; Scharf,S.J.; Higuchi,R.; Horn,G.T.; Mullis,K.B.; Erlich, H.A. *Science* **1988**, *239*:487-491.
3.Adams,R.L.P.; Knowler,J.T.; Leader,D.P. *The Biochemistry of the Nucleic Acids*; Chapman and Hall: New York, 1986, pp 146-161.
4.Leob,L.A.; Kunkle,T.A. *Ann. Rev. Biochem.* 1982, 52:429-457.
5.Muzyczka,N.; Poland,R.L.; Bessman,M.J. *J. Biol. Chem.* **1972**, *247*:7116-7122.
6.Fersht,A.R.; Knill-Jones,J.W. *J. Mol. Biol.* **1983**, *165*:669-682.
7.Sinha,N.K. *Proc. Natl. Acad. Sci. USA* **1987**, *84*:915-919.
8.Reyland,M.E.; Lehman,I.R.; Loeb,L.A. *J. Biol. Chem.* **1988**, *263*:6518-6524.
9.Kunkle,T.A.; Beckman,R.A.; Loeb,L.A. *J. Biol. Chem.* **1986**, *261*:13610-13616.
10.Bernad,A.; Blanco,L.; Lazaro,J.M.; Martin,G.; Salas,M.A. *Cell.* **1989**, *59*:219-228.

11.Loeb,L.A.; Reyland,M.E. *Nucleic Acids and Molecular Biology*; F. Eckstein and D.M.J. Lilley, Eds: Springer-Verlag Berlin Heidelberg, 1987, Vol 1. pp 157-173

12.Tindall,K.R.; Kunkle,T.A. *Biochemistry* **1988**, *27*:6008-6013.

13.Keohavong,P.; Thilly,W.G. *Proc. Natl. Acad. Sci. USA* **1989**, *86*:9253-9257.

14.Eckert,K.A.; Kunkle,T.A. *Nucl. Acids Res.* **1990**, *18*:3739-3744.

15.Scharf,S.; Horn,G.T.; Erlich,H.A. *Science* **1986**, *233*:1076-1078.

16.Felts,K.; Weiner,M.; Braman,J. *Stratagies* **1992**, in press.

17.Manam,S.; Nichols,W.W. *Analytical Biochem.* **1991**, *199*:106-111.

18.Lacy,M.J.; McNeil,M.E.; Roth,M.E.; Kranz,D.M. Proc. Natl. Acad. Sci. USA 1989, *86*:1023-1026.

19.Frohman,M.A.; Dush,M.K.; Martin,G.R. *Proc. Natl. Acad. Sci. USA* **1988**, *85*:8998-9002.

20.Li.H.; Cui,X.; Arnheim,N. *Proc. Natl. Acad. Sci. USA* **1990**, *87*:4580-4584.

21.Jeffreys,A.J.; Neuman,R.; Wilson,V. *Cell* **1990**, *60*:473-485.

22.Ruano,G.; Kidd,K.K.; Stephens,J.C. *Proc. Natl. Acad. Sci. USA* **1990**, *87*:6296-6300.

23.Lundberg,K.S.; Shoemaker,D.D.; Adams,M.W.W.; Short,J.M.; Sorge,J.A.; Mathur,E.J. *Gene* **1991**, *108*:1-6.

24.Bergseid,M.; Scott,B.; Mathur,S.; Nielson,K.; Shoemaker,D.; Mathur,E.J. *Stratagies* **1991**, *4*:34-35.

25.Fiala,G.; Stetter,K.O. *Arch. Microbiol.* **1986**, *145*:56-   61.

26.Mathur,E.J.; Adams,M.W.W.; Callen,W.N.; Cline,J.M. *Nucl. Acids Res.* **1991**, *19*:6952.

27.Kohler,S.W.; Provost,G.S.;Fiek,A.; Kretz,P.L.; Bullock,W.O.; Sorge,J.A.; Putman,D.L. and Short,J.M. *Proc. Natl. Acad. Sci. USA* **1991**, *88*:7958-7962.

28.Gordon,A.J.E.; Burns,P.A.; Fix,D.F.; Yatagai,F.; Allen,F.L.; Horsfall,M.J.; Halliday,J.A.; Gray,J.; Bernelot-Moens,C.; Glickman,B.W. *J. Mol. Biol.* **1988**, *200*:239-251.

29.Krishnan,B.R.; Blakesley,R.W.; Berg,D.E. *Nucl. Acids Res.* **1991**, *19*:1153

30.Murray,V. *Nucl. Acids Res.* **1989**, *17*:8889.

31.Barany,F. Proc. Natl. Acad. Sci. USA **1991**, *88*:189-193.

32.Hampl,H.; Marshall,R.A.; Perko,T.; Solomon,N. *PCR Topics: Usage of the Polymerase Chain Reaction in Genetics and Infectious Diseases*; Rolfs,A.; Schumacher,M.; and Marx,P.,Eds: Springer-Verlag Berlin Heidelberg, 1991, pp.15-22

33.Mullis,K.B. *Scientific American* **1990**, *4*:56-65.

RECEIVED January 15, 1992

INDEXES

Author Index

Affiliation Index

Subject Index

Production: C. Buzzell-Martin
Indexing: Deborah H. Steiner
Acquisition: Barbara C. Tansill
Cover design: Alan Kahan

Printed and bound by Maple Press, York, PA